TRIGONOMETRY

Cameron B. Douthitt
Dean, Research, Planning and Development
Alvin Community College

Joe A. McMillian
Professor of Mathematics, Division Chairman
Mathematics, Science, and Engineering
North Harris County College

McGraw-Hill Book Company
New York / St. Louis / San Francisco
Auckland / Bogotá / Düsseldorf
Johannesburg / London / Madrid
Mexico / Montreal / New Delhi
Panama / Paris / São Paulo
Singapore / Sydney / Tokyo / Toronto

TRIGONOMETRY

1 2 3 4 5 6 7 8 9 0 K P K P 7 8 3 2 1 0 9 8 7

This book was set in Times Roman by Black Dot, Inc.
The editors were A. Anthony Arthur and Frances A. Neal;
the designers were J. Paul Kirouac and Craigwood Phillips, A Good Thing, Inc.;
the production supervisor was Robert C. Pedersen.
Kingsport Press, Inc. was printer and binder.

Library of Congress Cataloging in Publication Data

Douthitt, Cameron B
    Trigonometry.

    Includes index.
    1. Trigonometry.   I. McMillian, Joe A., joint
author. II. Title.
QA531.D68   516'.24   76-44544
ISBN 0-07-017670-1

CONTENTS

C

PREFACE

This text is designed to provide a flexible approach to the study of trigonometry. The structure of trigonometry is maintained while utilizing a presentation which is easily understood by students with nonhomogeneous backgrounds. A competency in basic algebra is the only prerequisite; however, reviews of algebra and logarithms are provided in the Appendix.

Functions of angles in degree measure and radian measure are defined in the first two chapters. An early introduction to right triangles and applications is given and includes some use of metric measure.

A great deal of flexibility is allowed by the book's design, and no loss of continuity will arise from various orderings of the chapters. Several possible arrangements are suggested in the teacher's manual. Thus, the text is well suited for a one-semester course for any type of trigonometry student.

The book includes many unusual features. Each chapter begins with a list of measurable objectives so that the student knows specifically what must be accomplished. The objectives, correlated to a particular section by the numbers in parentheses, enable the instructor to teach for any level of mastery. The text also contains a large selection of problems at different levels of difficulty. The degree of detail in the examples and proofs enables the student to obtain reinforcement on a step-by-step basis.

The chapter on graphing has some unique characteristics which enable the student to sketch easily both basic and difficult graphs. Changes in amplitudes, periods, and positional shifts are easily calculated through the use of tables of reference points.

The instructor's manual contains teaching suggestions, two tests for each chapter, comprehensive exams, and solutions to all tests. The text is well suited to a self-paced, laboratory course where individual instruction is provided by an instructor and tutors.

The bock has been taught successfully in courses of both lecture and laboratory type at North Harris County College, Alvin Community College, and The University of Houston (Technical Mathematics Department). Student responses and evaluations indicate the efficacy of the pedagogical methods to which the text lends itself.

The authors are grateful to Kathleen Baldwin, Teresa Gordy, Barbara Kincade, Mary Miller, Teresa Miller, Joyce Osteen, and Sandy Tate for their assistance in the preparation of the manuscript.

Cameron B. Douthitt
Joe A. McMillian

TO THE STUDENT

S

You are about to begin the study of a branch of mathematics that has broad applications in surveying, drafting, astronomy, navigation, physics, and engineering. Trigonometry means triangle measurement. It is one of the oldest areas of mathematics and remains both interesting and enjoyable.

Trigonometry, like all branches of mathematics, should be studied as a game. It has players, rules, and strategy. In order to "win" at trigonometry you must develop a winning strategy that obeys the rules. This text is designed to help you become a "winner."

Each chapter begins with a set of objectives. These objectives should serve as a check list for the skills you develop. Each objective is followed by numbers that tell you which sections in the chapter are related to that particular objective.

Each section explains how the rules of the game apply to the particular topic under discussion. Several examples are given prior to each exercise, so study these examples carefully and pay close attention to the definitions and rules that are given. Understanding the definitions and rules is essential before you can learn to develop strategies that will help you to "win."

At the end of each chapter you will find a review. You should try to work these problems as if they were a test. If you find you are having trouble mastering a particular objective, you may wish to restudy the appropriate section. Answers to all exercises and reviews are given so that you will have immediate results on your progress.

Remember, one "learns" mathematics by "doing" mathematics, good luck and enjoy your game!

Cameron B. Douthitt
Joe A. McMillian

Angles and Measurement

1

1/Angles and Measurement

1-1 Introduction Trigonometry is a part of mathematics that includes a group of functions that are closely related to angles and triangles. Before defining these "trigonometric" functions, we begin our study with a careful examination of angles and their measurement.

1-2 Angles in Trigonometry In plane geometry, an angle is defined as the union of two rays beginning from a common point. In trigonometry we define an *angle* to be the rotation required to move one ray (initial side) to the position of another ray (terminal side) where the rays have a common endpoint called the "vertex" or "origin."

Figure 1-2.1

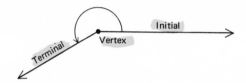

The angle is positive if the rotation is counterclockwise and negative if the rotation is clockwise. A curved arrow indicates the direction of the rotation. Small Greek letters are often used to represent angles. Some of the most freqently used Greek letters are α (alpha), β (beta), γ (gamma), ρ (rho), and θ (theta). Some angles are sketched in Figures 1-2.2, 1-2.3, and 1-2.4.

Figure 1-2.2

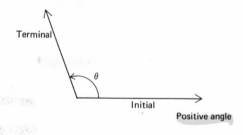

Figure 1-2.3

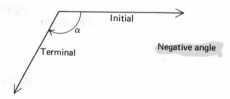

Figure 1-2.4

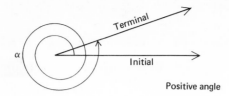

Positive angle

1-3   Angle Measure

To measure the amount of rotation of an angle, we use the *revolution*, the *degree*, or the *radian*.

One *revolution* is defined as one complete rotation where the terminal side is rotated, clockwise or counterclockwise, one time to coincide with the initial side.

Example 1-3.1   Sketch a positive angle of one revolution.

Solution

Figure 1-3.1

One *degree* is defined as the measure of rotation of an angle which equals $\frac{1}{360}$ of one revolution.   Hence, $1° = \frac{1}{360}$ revolution and $360° = 1$ revolution.   One *minute* (1′) equals $\frac{1}{60}$ of 1° and one *second* (1″) equals $\frac{1}{60}$ of 1′.

A *radian* is the measure of a central angle of a circle which intercepts an arc equal in length to the radius of the circle.

Example 1-3.2   Sketch an angle of −2.5 revolutions.

Solution

Figure 1-3.2

Example 1-3.3   Sketch an angle of 210°.

Solution

Figure 1-3.3

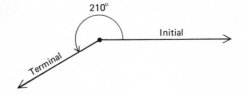

Example 1-3.4   Sketch an angle of 2.3 radians.

Solution   Since the measure of a right angle is 1.57 $\left(\dfrac{\pi}{2}\right)$ radians and the measure of a straight angle is 3.14 (π) radians, we can approximate the position of the terminal side of 2.3 radians.

Figure 1-3.4

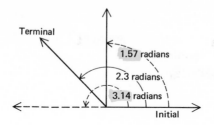

Exercise 1-3.1   Sketch the following angles and label the initial and terminal sides.   Draw a curved arrow to indicate positive or negative direction.

1   75°   **2**   110°   **3**   −180°

4   −470°   **5**   120°   **6**   −90°

7   105°   **8**   −278°   **9**   200°

1-4 Conversions of Angle Measure   From the definition we see that

$$1 \text{ revolution} = 2\pi \text{ radians} = 360°$$

Hence, π radians = 180°. This is a most important relationship, and we can use it to convert from radian measure to degree measure, or vice versa. Whenever angle measure is given in this book, it will be assumed to be radian measure unless the degree symbol is indicated; that is, 1 means 1 radian, 5° means 5 degrees. Also, 3.14 will be used for π in all calculations.

Since $\pi = 180°$, we see that

$$1 \text{ radian} = \frac{180°}{\pi} = 57°18' \text{ approximately}$$

$$1 \text{ degree} = \frac{\pi}{180°} = 0.01745 \text{ approximately}$$

In converting from one unit of measure to another, there is a standard procedure which is used extensively throughout science and mathematics. This procedure is based on the fact that any number multiplied by the number one remains unchanged; that is, $x \cdot 1 = x$ for any value x. Now let's apply this procedure to convert degrees to radians and vice versa.

Suppose we wish to convert $90°$ to radian measure:

$$90 \text{ degrees} = 90 \text{ degrees} \cdot 1$$

$$= 90 \text{ degrees} \cdot \frac{\pi \text{ radians}}{\pi \text{ radians}}$$

$$= 90 \text{ degrees} \cdot \frac{\pi \text{ radians}}{180 \text{ degrees}}$$

$$= \frac{90 \text{ degrees}}{1} \cdot \frac{\pi \text{ radians}}{180 \text{ degrees}}$$

$$= \frac{\pi \text{ radians}}{2}$$

$$= \frac{\pi}{2}$$

To convert $\frac{\pi}{10}$ to degrees, we have

$$\frac{\pi}{10} \text{ radians} = \frac{\pi}{10} \text{ radians} \cdot 1$$

$$= \frac{\pi}{10} \text{ radians} \cdot \frac{180 \text{ degrees}}{180 \text{ degrees}}$$

$$= \frac{\pi}{10} \text{ radians} \cdot \frac{180 \text{ degrees}}{\pi \text{ radians}}$$

$$= 18 \text{ degrees}$$

Recalling that $1° = 0.01745$ radian approximately, we may use this as an alternate conversion factor from radians to degrees and vice versa.

To convert 90° to radians, we write

$$90 \text{ degrees} = 90 \text{ degrees} \cdot 1$$

$$= 90 \text{ degrees} \cdot \frac{1 \text{ degree}}{1 \text{ degree}}$$

$$= 90 \text{ degrees} \cdot \frac{0.01745 \text{ radian}}{1 \text{ degree}}$$

$$= 1.57050 \text{ radians approximately}$$

To convert $\frac{\pi}{10}$ radians to degrees, we write

$$\frac{\pi}{10} \text{ radians} = \frac{\pi}{10} \text{ radians} \cdot 1$$

$$= \frac{\pi}{10} \text{ radians} \cdot \frac{1 \text{ degree}}{1 \text{ degree}}$$

$$= \frac{3.14}{10} \text{ radians} \cdot \frac{1 \text{ degree}}{0.01745 \text{ radian}}$$

$$= 17.99 \text{ degrees approximately}$$

Note that when the exact relationship $\pi = 180°$ is used, exact answers of $90° = \frac{\pi}{2}$ and $\frac{\pi}{10} = 18°$ are obtained. $1° = 0.01745$ radian is an approximation; so, whenever this conversion factor is used, approximate answers are obtained. That is, $90° = 1.5705$ radians approximately and $\frac{\pi}{10} = 17.99°$ approximately.

The following examples further illustrate the technique for converting units of measure.

Example 1-4.1   Convert 30° to radians.

Solution $$30° = 30° \cdot 1$$

$$= 30° \cdot \frac{\pi \text{ radians}}{\pi \text{ radians}}$$

$$= 30° \cdot \frac{\pi \text{ radians}}{180 \text{ degrees}}$$

$$= \frac{\pi}{6} \text{ radians}$$

Example 1-4.2 Convert $\dfrac{3\pi}{4}$ to degrees using π radians $= 180°$.

Solution $\dfrac{3\pi}{4} = \dfrac{3\pi}{4}$ radians $\cdot$ 1

$\qquad = \dfrac{3\pi}{4}$ radians $\cdot \dfrac{180 \text{ degrees}}{180 \text{ degrees}}$

$\qquad = \dfrac{3\pi}{4}$ radians $\cdot \dfrac{180 \text{ degrees}}{\pi \text{ radians}}$

$\qquad = 135 \text{ degrees}$

Example 1-4.3 Convert $\dfrac{3\pi}{4}$ to degrees using 1 degree $= .01745$ radian.

Solution $\dfrac{3\pi}{4} = \dfrac{3\pi}{4}$ radians $\cdot$ 1

$\qquad = \dfrac{3\pi}{4}$ radians $\cdot \dfrac{1 \text{ degree}}{1 \text{ degree}}$

$\qquad = \dfrac{3(3.14)}{4}$ radians $\cdot \dfrac{1 \text{ degree}}{0.01745 \text{ radian}}$

$\qquad = 134.96 \text{ degrees approximately}$

RULE To convert from degrees to radians, multiply degree measure by $\dfrac{\pi}{180°}$.

To convert from radians to degrees, multiply radian measure by $\dfrac{180°}{\pi}$.

Exercise 1-4.1 Use 3.14 as an approximation for π.

1 Convert the following angles to radian measure:

    (a) 45°   (b) $-135°$ (c) 210°   (d) 105°

    (e) 330°   (f) $-315°$ (g) 55°   (h) $-200°$

2 Convert the following angles to degree measure:

    (a) $\tfrac{3}{2}\pi$ (b) $-\tfrac{2}{3}\pi$ (c) $\tfrac{5}{2}\pi$ (d) 3.9π

    (e) -10π (f) $\tfrac{4}{3}\pi$ (g) $-\tfrac{7}{4}\pi$ (h) -6π

1-5 Standard Position of an Angle

An angle is defined to be in *standard position* whenever the vertex coincides with the origin of a rectangular coordinate system and the initial side lies along the positive x axis.

Example 1-5.1 Sketch $60°$ in standard position.

Solution

Figure 1-5.1

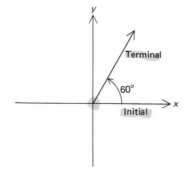

Example 1-5.2 Sketch $-210°$ in standard position.

Solution

Figure 1-5.2

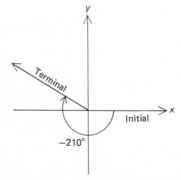

Angles in standard position with the same terminal side are said to be *coterminal*. Hence, $-45°, 315°,$ and $-405°$ are coterminal angles. An angle with its terminal side along one of the axes is called a *quadrantal* angle.

Example 1-5.3 Sketch the coterminal angles $-45°$, $315°$, and $-405°$.

Solution

Figure 1-5.3

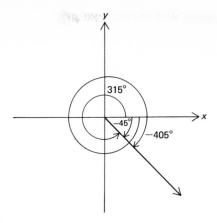

Example 1-5.4 Sketch the coterminal quadrantal angles of $90°$, $-270°$, and $450°$.

Solution

Figure 1-5.4

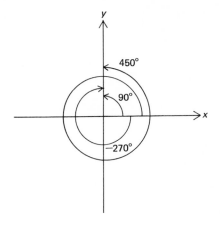

Exercise 1-5.1 **1** Sketch the following angles in standard position:

(a) 25°   (b) −25°   (c) 160°   (d) π (e) −4

2 Find two angles, one positive and one negative, coterminal with each angle in the previous problem.

1-6 Finding the Complement and Supplement of an Angle

Angles are said to be *complementary* if their sum is 90° $\left(\dfrac{\pi}{2}\right)$. Angles are *supplementary* if their sum is 180° (π). The following examples illustrate how to find the complement and supplement of a given angle.

Example 1-6.1   Find the complement of 68°.

Solution
$$x + 68° = 90°$$
$$x = 22°$$

Example 1-6.2   Find the supplement of 68°.

Solution
$$x + 68° = 180°$$
$$x = 112°$$

Example 1-6.3   Find the complement of −71°27′.

Solution
$$x + (-71°27') = 90°$$
$$x = 90° + 71°27'$$
$$x = 161°27'$$

Example 1-6.4   Find the supplement of −98°31′20″.

Solution
$$x + (-98°31'20'') = 180°$$
$$x = 278°31'20''$$

Example 1-6.5   Find the complement of $\dfrac{2\pi}{3}$.

Solution
$$x + \frac{2\pi}{3} = \frac{\pi}{2}$$
$$x = \frac{\pi}{2} - \frac{2\pi}{3}$$
$$x = \frac{3\pi}{6} - \frac{4\pi}{6}$$
$$x = -\frac{\pi}{6}$$

Example 1-6.6   Find the supplement of $-\dfrac{7\pi}{8}$.

Solution $x + -\dfrac{7\pi}{8} = \pi$

$x = \pi + \dfrac{7\pi}{8}$

$x = \dfrac{15\pi}{8}$

Exercise 1-6.1   **1** Find the complement of each of the following angles:

(a) 27°55'   (b) 88°9'   (c) $\dfrac{\pi}{4}$ (d) $-\dfrac{2\pi}{3}$

(e) 33°45'27"   (f) 65°13'4"   (g) 210°

2 Find the supplement of each of the following angles:

(a) 129°37'   (b) −31°15'32"   (c) $\dfrac{3\pi}{5}$

(d) 111°11'11"   (e) 347°   (f) 270°5'5"

Review Exercise   **1** Sketch the following angles. Label the initial and terminal sides.

(a) 45°   (b) 60°   (c) 85°   (d) −20°

(e) −120°   (f) −90°   (g) −135°   (h) −220°

(i) −315°   (j) −390°

2 Convert each angle to radian measure.

(*a*) 40° (*b*) 60° (*c*) −45° (*d*) 120°

(*e*) 190° (*f*) −95° (*g*) −425° (*h*) 288°

(*i*) 130° (*j*) −54°

3 Convert each angle to degree measure.

(*a*) $\dfrac{7\pi}{4}$ (*b*) $\dfrac{\pi}{3}$ (*c*) $\dfrac{\pi}{6}$ (*d*) $-\dfrac{\pi}{5}$

(*e*) $-\dfrac{10\pi}{7}$ (*f*) -16π (*g*) $\dfrac{4\pi}{3}$ (*h*) $\dfrac{6\pi}{7}$

(*i*) $-\dfrac{\pi}{9}$ (*j*) $-\dfrac{5\pi}{6}$

4 Sketch the following angles in standard position:

(*a*) 80° (*b*) −120° (*c*) −390° (*d*) 495°

5 Write two angles, one positive and one negative, that are coterminal with the following angles:

(*a*) 30° (*b*) 60° (*c*) 120° (*d*) −180°

(*e*) −18° (*f*) $-\dfrac{\pi}{3}$ (*g*) $\dfrac{\pi}{2}$ (*h*) $\dfrac{\pi}{6}$

(*i*) $-\dfrac{5\pi}{6}$ (*j*) $-\dfrac{5\pi}{9}$

6 Write the complement of the following angles:

(a) 30°   (b) 60°   (c) 45°   (d) 50°

(e) 88°   (f) 100°   (g) $\dfrac{\pi}{6}$ (h) $\dfrac{5\pi}{6}$

(i) $-\dfrac{3\pi}{4}$ (j) $-\dfrac{7\pi}{8}$

7 Write the supplement of the following angles:

(a) 30°   (b) 60°   (c) 45°   (d) 50°

(e) 88°   (f) 100°   (g) $\dfrac{\pi}{6}$ (h) $\dfrac{5\pi}{6}$

(i) $-\dfrac{3\pi}{4}$ (j) $-\dfrac{7\pi}{8}$

Answers

Exercise 1-3.1

1

2

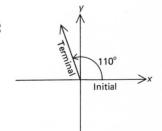

3

4

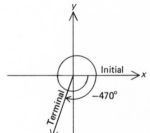

5

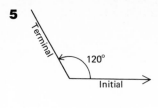

120°

6

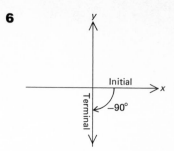

7

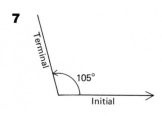

105°

8

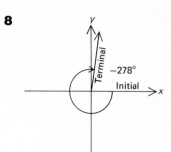

9

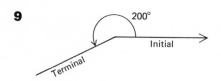

200°

Exercise 1-4.1

1 (a) $\dfrac{\pi}{4}$ (b) $\dfrac{-3\pi}{4}$ (c) $\dfrac{7\pi}{6}$ (d) $105° = \dfrac{7\pi}{12}$

(e) $\dfrac{11\pi}{6}$ (f) $-315° = \dfrac{-7\pi}{4}$ (g) $\dfrac{11\pi}{36}$ (h) $\dfrac{-10\pi}{9}$

2 (a) $270°$ (b) $-120°$ (c) $450°$ (d) $702°$

(e) $-1800°$ (f) $240°$ (g) $-315°$ (h) $-1080°$

Exercise 1-5.1

1 (a)

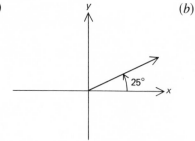

25°

(b)

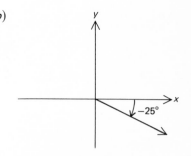

−25°

(c)

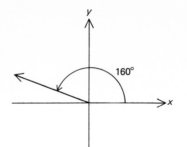

(d)

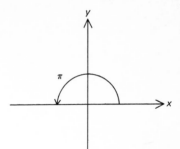

(e)

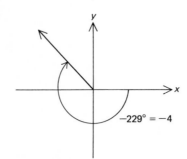

2 (a) $-335°$, $385°$ (b) $335°$, $-385°$
(c) $-200°$, $520°$ (d) $-180°$, $540°$
(e) $131°$, $-589°$

Exercise 1-6.1 **1** (a) $62°5'$ (b) $1°51'$ (c) $\dfrac{\pi}{4}$ (d) $\dfrac{7\pi}{6}$
(e) $56°14'33''$ (f) $24°46'56''$ (g) $-120°$

2 (a) $50°23'$ (b) $211°15'32''$ (c) $\dfrac{2\pi}{5}$
(d) $68°48'49''$ (e) $-167°$ (f) $-90°5'5''$

Review Exercise **1** (a)

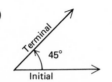

(b)

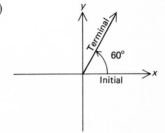

(c)

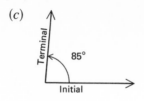

(d)

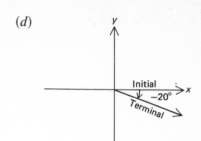

(e)

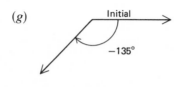

(f)

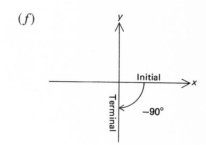

(g)

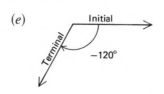

(h)

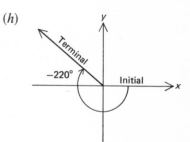

(i)

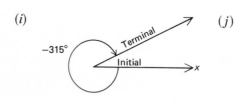

(j)

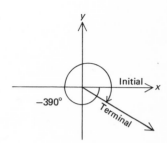

2 (a) $\dfrac{2\pi}{9}$ (b) $\dfrac{\pi}{3}$ (c) $-\dfrac{\pi}{4}$ (d) $\dfrac{2\pi}{3}$

(e) $\dfrac{19\pi}{18}$ (f) $\dfrac{-19\pi}{36}$ (g) $-\dfrac{85\pi}{36}$ (h) $\dfrac{8\pi}{5}$

(i) $\dfrac{13\pi}{18}$ (j) $\dfrac{-3\pi}{10}$

3 (*a*) 315°   (*b*) 60°   (*c*) 30°   (*d*) −36°
  (*e*) −257$\frac{1}{7}$°   (*f*) −2880°   (*g*) 240°   (*h*) 154$\frac{2}{7}$°
  (*i*) −20°   (*j*) −150°

4 (*a*)   (*b*)

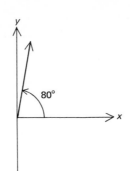   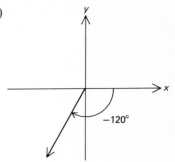

  (*c*)   (*d*)

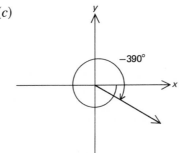   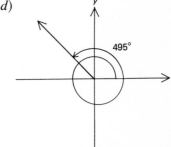

5 (*a*) 390°, −330°   (*b*) 420°, −300°   (*c*) 480°, −240°

  (*d*) 180°, −540°   (*e*) 342°, −378°   (*f*) $\dfrac{5\pi}{3}, \dfrac{-7\pi}{3}$

  (*g*) $\dfrac{5\pi}{2}, \dfrac{-3\pi}{2}$ (*h*) $\dfrac{13\pi}{6}, \dfrac{-11\pi}{6}$ (*i*) $-\dfrac{17\pi}{6}, \dfrac{7\pi}{6}$

  (*j*) $\dfrac{13\pi}{9}, \dfrac{-23\pi}{9}$

6 (*a*) 60°   (*b*) 30°   (*c*) 45°   (*d*) 40°

  (*e*) 2°   (*f*) −10°   (*g*) $\dfrac{\pi}{3}$ (*h*) $-\dfrac{\pi}{3}$

  (*i*) $\dfrac{5\pi}{4}$ (*j*) $\dfrac{11\pi}{8}$

7 (*a*) 150°   (*b*) 120°   (*c*) 135°
  (*d*) 130°   (*e*) 92°   (*f*) 80°

  (*g*) $\dfrac{5\pi}{6}$ or 150°   (*h*) $\dfrac{\pi}{6}$ or 30°   (*i*) $\dfrac{7\pi}{4}$ or 315°

  (*j*) $\dfrac{15\pi}{8}$

Trigonometric Functions

2

2/Trigonometric Functions

2-1 Definitions of Trigonometric Functions of Angles

Consider a positive angle θ in standard position in the xy plane. Then there is some number r which represents the length from $(0,0)$ to $P(x,y)$. Since a right triangle can be formed using the x axis and the terminal side of θ as the hypotenuse, we observe that $r^2 = x^2 + y^2$ by the pythagorean theorem. Only the positive value of r is used since it represents length; hence $r = \sqrt{x^2 + y^2}$.

Figure 2-1.1

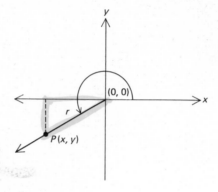

Example 2-1.1 If θ is an angle in standard position and $P(3,-4)$ is on its terminal side, find r.

Solution $\qquad r^2 = x^2 + y^2 \qquad\qquad x = 3,\ y = -4$

$\qquad\qquad r^2 = 3^2 + (-4)^2$

$\qquad\qquad r^2 = 25$

$\qquad\qquad r = 5$

Figure 2-1.2

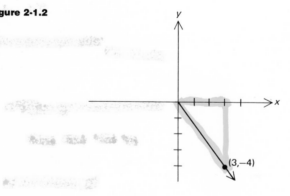

Example 2-1.2   If θ is an angle in standard position and $P(x, \sqrt{3})$ is on its terminal side with $r = 7$, find x.

Solution
$$7 = \sqrt{x^2 + (\sqrt{3})^2}$$
$$49 = x^2 + 3$$
$$46 = x^2$$
$$\pm\sqrt{46} = x$$

Note that this means that P can be in quadrant I or II.

Figure 2-1.3

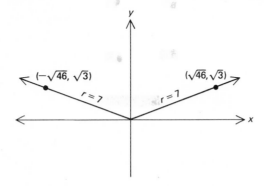

If the point $P(x,y)$ is on the terminal side of an angle θ in standard position and r is the distance from the origin to P, the trigonometric functions of θ are sine, cosine, tangent, cotangent, secant, and cosecant, as defined below. Their respective abbreviations are sin, cos, tan, cot, sec, and csc.

$IN = \dfrac{OPP}{HYP} \dfrac{y}{r}$ $CSC = \dfrac{HYP}{OPP} \dfrac{r}{y}$ $\sin \theta = \dfrac{y}{r}$ $r > 0$ $\cos \theta = \dfrac{x}{r}$ $r > 0$

$OS = \dfrac{ADJ}{HYP} \dfrac{x}{r}$ $SEC = \dfrac{HYP}{ADJ} \dfrac{r}{x}$ $\tan \theta = \dfrac{y}{x}$ $x \ne 0$ $\cot \theta = \dfrac{x}{y}$ $y \ne 0$

$AN = \dfrac{OPP}{ADJ} \dfrac{y}{x}$ $COT = \dfrac{ADJ}{OPP} \dfrac{x}{y}$ $\sec \theta = \dfrac{r}{x}$ $x \ne 0, r > 0$ $\csc \theta = \dfrac{r}{y}$ $y \ne 0, r > 0$

If two of the three values for x, y, and r are known, then the values of the six trigonometric functions can be found.

It is important to know the signs of the trigonometric functions of an angle in any quadrant. From the definitions of the functions and the signs for x, y, and r in the various quadrants, the values of the functions are positive or negative, as shown in Figure 2-1.4.

Figure 2-1.4

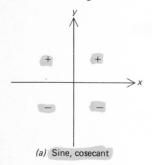

(a) Sine, cosecant

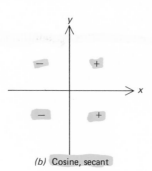

(b) Cosine, secant

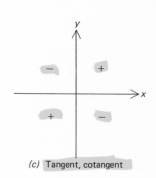

(c) Tangent, cotangent

Example 2-1.3   Write the values of the trigonometric functions of angle θ in standard position if $P(3,2)$ is on the terminal side of θ.

Solution   If $P(3,2)$ is on the terminal side of θ, then $r = \sqrt{3^2 + 2^2} = \sqrt{13}$, and hence

$$\sin \theta = \frac{2}{\sqrt{13}} = \frac{2\sqrt{13}}{13} \qquad \tan \theta = \frac{2}{3} \qquad \sec \theta = \frac{\sqrt{13}}{3}$$

$$\cos \theta = \frac{3}{\sqrt{13}} = \frac{3\sqrt{13}}{13} \qquad \cot \theta = \frac{3}{2} \qquad \csc \theta = \frac{\sqrt{13}}{2}$$

Figure 2-1.5

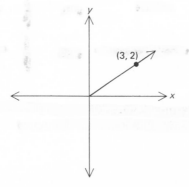

(3, 2)

Example 2-1.4 If $P(x,-4)$ is on the terminal side of θ and $r = 5$, write the values of the trigonometric functions of θ.

Solution
$$5 = \sqrt{x^2 + (-4)^2}$$
$$= \sqrt{x^2 + 16}$$
$$25 = x^2 + 16$$
$$9 = x^2$$
$$\pm 3 = x$$

$$\sin \theta = -\tfrac{4}{5} \qquad \tan \theta = \pm\tfrac{4}{3} \qquad \sec \theta = \pm\tfrac{5}{3}$$
$$\cos \theta = \pm\tfrac{3}{5} \qquad \cot \theta = \pm\tfrac{3}{4} \qquad \csc \theta = -\tfrac{5}{4}$$

Figure 2-1.6

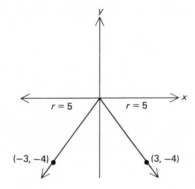

Example 2-1.5 If $\cos \theta = 0.6$ and θ is in quadrant IV, find the other trigonometric ratios.

Solution Since $0.6 = \tfrac{6}{10} = \tfrac{3}{5}$, $\cos \theta = \tfrac{3}{5} = \dfrac{x}{r}$, so $x = 3$, $r = 5$, and hence $y = -4$.

$$\sin \theta = -\tfrac{4}{5} \qquad \tan \theta = -\tfrac{4}{3} \qquad \sec \theta = \tfrac{3}{5}$$
$$\cos \theta = \tfrac{3}{5} \qquad \cot \theta = -\tfrac{3}{4} \qquad \csc \theta = -\tfrac{5}{4}$$

Figure 2-1.7

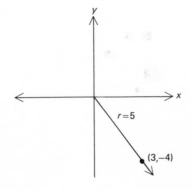

Exercise 2-1.1 **1** Suppose $P(s,t)$ is on the terminal side of angle θ in standard position and r is the distance from $(0,0)$ to P. Write the six trigonometric ratios of θ.

2 Write the values of the six trigonometric functions of the angle θ in standard position if:
(*a*) $(12,5)$ is on the terminal side of θ.
(*b*) $(-12,-5)$ is on the terminal side of θ.
(*c*) $(-4,-3)$ is on the terminal side of θ.
(*d*) $(\sqrt{3},1)$ is on the terminal side of θ.
(*e*) $(x,5)$ is on the terminal side of θ and $r = 13$.

3 Find the values of the remaining trigonometric functions of angle α in standard position.
(*a*) $\csc \alpha = -\frac{5}{3}$, $\cos \alpha > 0$
(*b*) $\cot \alpha = 3$, $\sec \alpha < 0$
(*c*) $\sec \alpha = \frac{13}{5}$, $\sin \alpha < 0$
(*d*) $\tan \alpha = -\frac{1}{3}$, $\csc \alpha < 0$
(*e*) $\cot \alpha = -3$, α in quadrant II
(*f*) $\sin \alpha = -\frac{1}{2}$, α in quadrant III
(*g*) $\cos \alpha = -0.25$, α in quadrant II
(*h*) $\tan \alpha = 0.8$, α in quadrant III

2-2 Functions of Special Angles Special right triangles having angles of 30°, 60°, or 45° are studied in plane geometry. If a base angle of 30°, 45°, or 60° is placed in standard position, as shown in Figure 2-2.1, exact values of the trigonometric functions of the angle can be found.

Figure 2-2.1

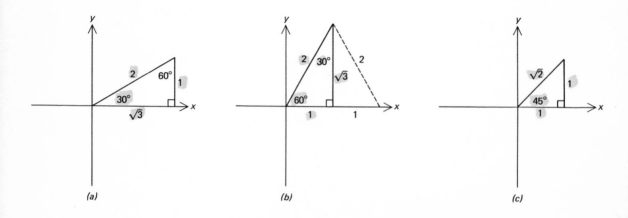

(a)   (b)   (c)

The definitions of the trigonometric functions can be employed to find the values for the trigonometric functions of each of these angles, as given in Table 2-2.1.

Table 2-2.1

θ	$\sin \theta$	$\cos \theta$	$\tan \theta$	$\cot \theta$	$\sec \theta$	$\csc \theta$
$30° \left(\dfrac{\pi}{6}\right)$	$\dfrac{1}{2}$	$\dfrac{\sqrt{3}}{2}$	$\dfrac{1}{\sqrt{3}} = \dfrac{\sqrt{3}}{3}$	$\sqrt{3}$	$\dfrac{2}{\sqrt{3}} = \dfrac{2\sqrt{3}}{3}$	2
$45° \left(\dfrac{\pi}{4}\right)$	$\dfrac{1}{\sqrt{2}} = \dfrac{\sqrt{2}}{2}$	$\dfrac{1}{\sqrt{2}} = \dfrac{\sqrt{2}}{2}$	1	1	$\sqrt{2}$	$\sqrt{2}$
$60° \left(\dfrac{\pi}{3}\right)$	$\dfrac{\sqrt{3}}{2}$	$\dfrac{1}{2}$	$\sqrt{3}$	$\dfrac{1}{\sqrt{3}} = \dfrac{\sqrt{3}}{3}$	2	$\dfrac{2}{\sqrt{3}} = \dfrac{2\sqrt{3}}{3}$

A technique for remembering the functional values given in Table 2-2.1 is as follows:

For the functions of 45°, construct a right isosceles triangle with sides one unit in length (see Figure 2-2.1).

For the functions of 60° and 30°, construct an equilateral triangle with each side two units in length. Draw a perpendicular from one vertex to the opposite side to form two right triangles with sides 1 and $\sqrt{3}$ units long and a hypotenuse of 2 units (see Figure 2-2.1). The functional values of 30° and 60° may be read off these triangles.

The *reference angle* (often called the "related angle") for an angle θ in standard position is the angle α, $0° \leq \alpha \leq 90°$, such that one side of α is the terminal side of θ and the other side is the x axis with the vertex at the origin.

Example 2-2.1 Find the reference angle α for $\theta = 172°$.

Solution $180° - 172° = 8° = \alpha$

Figure 2-2.2

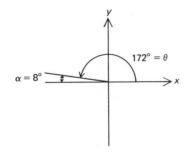

Example 2-2.2   Find the reference angle α for $\theta = 250°$.

Solution $180° + 70° = 250°$; thus $\alpha = 70°$.

Figure 2-2.3

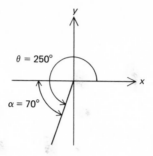

Example 2-2.3   Find the reference angle α for $\theta = 321°$.

Solution $\alpha = 360° - 321° = 39°$

Figure 2-2.4

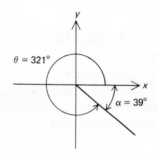

Observe that the exact value of the trigonometric functions can be found for any angle that has a reference angle of 30°, 45°, or 60°. The correct sign must be attached to the functional value. Three examples illustrate this idea.

Example 2-2.4   Find the values of the trigonometric functions of 150°.

Solution   The reference angle for 150° is 30°, and 150° is in quadrant II.

Figure 2-2.5

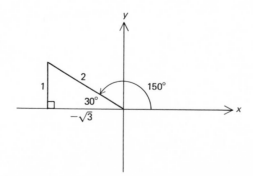

$$\sin 150° = \frac{y}{r} = \frac{1}{2} \qquad\qquad \tan 150° = \frac{y}{x} = -\frac{1}{\sqrt{3}} = -\frac{\sqrt{3}}{3}$$

$$\sec 150° = \frac{r}{x} = -\frac{2}{\sqrt{3}} = -\frac{2\sqrt{3}}{3} \qquad \cos 150° = \frac{x}{r} = -\frac{\sqrt{3}}{2}$$

$$\cot 150° = \frac{x}{y} = -\sqrt{3} \qquad\qquad \csc 150° = \frac{r}{y} = 2$$

Example 2-2.5   Find the values of the trigonometric functions of 315°.

Solution   The reference angle for 315° is 45°.

Figure 2-2.6

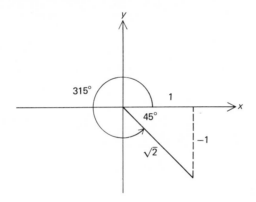

$$\sin 315° = -\frac{1}{\sqrt{2}} = -\frac{\sqrt{2}}{2} \qquad \tan 315° = -\frac{1}{1} = -1$$

$$\sec 315° = \frac{\sqrt{2}}{1} = \sqrt{2} \qquad \cos 315° = \frac{1}{\sqrt{2}} = \frac{\sqrt{2}}{2}$$

$$\cot 315° = \frac{1}{-1} = -1 \qquad \csc 315° = \frac{\sqrt{2}}{-1} = -\sqrt{2}$$

Example 2-2.6   Write the values of the remaining trigonometric functions when $\cos \theta = \frac{1}{2}$ and $\sin \theta < 0$.

Solution   Since $\cos \theta > 0$ and $\sin \theta < 0$, θ is in Q IV. Also $x = 1$, $r = 2$, and $y = -\sqrt{3}$. Hence, $\sin \theta = -\frac{\sqrt{3}}{2}$, $\tan \theta = -\sqrt{3}$, $\cot \theta = -\frac{\sqrt{3}}{3}$, $\sec \theta = 2$, and $\csc \theta = -\frac{2\sqrt{3}}{3}$.

Exercise 2-2.1 **1** Find the reference angle for each of the following angles in standard position:

(*a*) 172° (*b*) 131°17′ (*c*) 181°19′32″ (*d*) 219°

(*e*) −369°38′ (*f*) 321°16′16″ (*g*) $\dfrac{7\pi}{8}$ (*h*) $-\dfrac{4\pi}{5}$

2 Use the exact values of the trigonometric functions of 30°, 45°, and 60° to complete the following table:

θ	sin θ	cos θ	tan θ	cot θ	sec θ	csc θ
120°						
135°						
210°						
225°						
240°						
300°						
330°						

3 Determine the values of the trigonometric functions of each of the following angles in standard position:

(*a*) 480° (*b*) −210° (*c*) $-\dfrac{11\pi}{6}$

(*d*) −60° (*e*) 585° (*f*) $\dfrac{11\pi}{3}$

4 Write the values of the remaining functions when:

(*a*) sin $\theta = \tfrac{1}{2}$, cos $\theta < 0$

(*b*) tan $\theta = -3$, cos $\theta < 0$

(*c*) sec $\theta = \dfrac{2}{\sqrt{3}}$, sin $\theta < 0$

5 Verify the following statements:

(a) $\left(\sin\dfrac{2\pi}{3}\right)\left(\cos\dfrac{5\pi}{6}\right) - \left(\cos\dfrac{2\pi}{3}\right)\left(\sin\dfrac{5\pi}{6}\right) = -\dfrac{1}{2}$

(b) $\left(\cos\dfrac{5\pi}{6}\right)\left(\cos\dfrac{2\pi}{3}\right) + \left(\sin\dfrac{5\pi}{6}\right)\left(\sin\dfrac{2\pi}{3}\right) = \dfrac{\sqrt{3}}{2}$

(c) $\dfrac{\sin 30° - \sin 45°}{2\cos\dfrac{\pi}{3} + 3\tan 45°} = \dfrac{1 - \sqrt{2}}{8}$

(d) $1 + \tan^2 45° = 2$ **Remark** $\tan^2 \theta$ means $(\tan\theta)^2$.

2-3
Trigonometric
Functions of
Quadrantal
Angles

An angle that has its terminal side along the x or y axis—that is, 0°, 90°, 180°, 270°, 360° and their integral multiples—is a *quadrantal angle*. Using the definitions of the trigonometric functions and the appropriate values for x and y, we can write the values of the trigonometric functions of any quadrantal angle. Note that for 0°, $x = r$ and $y = 0$; for 90°, $x = 0$ and $y = r$; for 180°, $x = -r$ and $y = 0$; for 270°, $x = 0$ and $y = -r$. Since division by zero is undefined, this means that some functions are undefined for certain angles. Table 2-3.1 gives the values for quadrantal angles. The student should verify each.

Table 2-3.1

θ	$\sin\theta$	$\cos\theta$	$\tan\theta$	$\cot\theta$	$\sec\theta$	$\csc\theta$
0° (0)	0	1	0	Undef.	1	Undef.
90° $\left(\dfrac{\pi}{2}\right)$	1	0	Undef.	0	Undef.	1
180° (π)	0	−1	0	Undef.	−1	Undef.
270° $\left(\dfrac{3\pi}{2}\right)$	−1	0	Undef.	0	Undef.	−1

Exercise 2-3.1

1 Determine the six values of the trigonometric functions of each of the following angles in standard position:

(a) 720°   (b) 450°   (c) $-\tfrac{3}{2}\pi$ (d) $-\pi$

(e) -7π (f) $\tfrac{13}{2}\pi$ (g) $-\tfrac{2}{3}\pi$ (h) 630°

2 Verify the following:

(a) $\cos 0° + \sin 270° = 0$

(b) $\sin \dfrac{\pi}{2} \cos \dfrac{3\pi}{2} + \cos \dfrac{\pi}{2} \sin \dfrac{3\pi}{2} = 0$

(c) $\cos 2\pi \cos \pi + \sin 2\pi \sin \pi = -1$

(d) $\tan(-\pi) \cot \dfrac{\pi}{2} - \sec 2\pi \cos \pi = 1$

2-4 Values of Trigonometric Functions of Angles between 0° and 90°

It is now our task to find values of functions of any angle by means of a table that gives approximations for these values. Table C-1 (in Appendix C) is a standard table which gives the approximate value for each of the six trigonometric functions of an angle between 0° and 90°. The values are given to four places, and they have been rounded off. That is, if the fifth-place number was greater than or equal to 5, the fourth place was increased by 1; the fourth-place number remained the same if the fifth-place number was less than 5.

Although the exercise for this section deals with angles between 0° and 90°, we will see that, through the use of reference angles, the same table can be used to find values of functions for angles greater than 90° or less than 0°.

Notice that we find the value of a trigonometric function of an angle between 0° and 45° by locating the angle in the left-hand column and the function at the top of the page. The intersection of the row containing the given angle and the column for the required function produces the desired value. For example, to find cos 21°, we locate 21° in the left-hand column and the cosine column at the top of the page. The row containing 21° and the column containing cosine yield a value of 0.9336.

To find the value for a function of an angle between 45° and 90°, read the angle from the right-hand column and the function at the bottom of the page. For example, tan 72° = 3.0777.

You will also see that 0.9336 is the value of sin 69°. This illustrates an important property—that the six trigonometric functions occur in three pairs of cofunctions: $\sin A = \cos(90° - A)$, $\tan A = \cot(90° - A)$, and $\sec A = \csc(90° - A)$. These properties of cofunctions will be verified in Chapter 9.

Exercise 2-4.1 Use your table of values for the trigonometric functions to find the value of each of the following:

1 sin 65°

2 cos 31°10′

3 tan 67°30′

4 cot 15°

5 sec 59°

6 csc 45°40′

7 sec 50°10′

8 cot 32°40′

9 cot 47°20′

10 cos 5°20′

11 tan 3°

12 csc 9°

13 tan 88°

14 tan 9°

15 sec 90°

2-5 Functions of Any Angle

The obvious question one now asks is, What is the value of a function of an angle larger than 90° or smaller than 0°? To find such values, we recall that a function of any angle can be expressed as a function of its reference angle, as shown in Section 2-2. The value can then be read from a table, and the appropriate sign (+ or −) can be attached depending on the quadrant of the angle. For example, cos 150° can be found by observing that cos 150° = −cos 30° = −$\sqrt{3}$/2, as illustrated in Figure 2-5.1. To find sin 195°, observe Figure 2-5.2. Hence, sin 195° = −sin 15° = −0.2588.

Figure 2-5.1

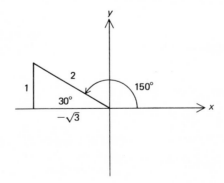

Figure 2-5.2

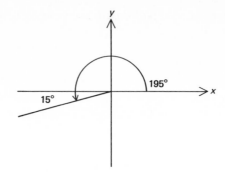

Exercise 2-5.1   Use Table C-1 to find the value of each of the following:

1   sin 200°   **2**   cos 172°20′   **3**   tan (−57°)

4   sec (−100°)   **5**   csc 321°51′   **6**   cot (−123°)

7   cos (−800°)   **8**   sec 450°   **9**   sin 230°30′

10   tan (−167°40′)

2-6   Interpolation   If we use methods that are beyond the scope of this course, it is possible to compute values of the trigonometric functions of any angle to the desired degree of accuracy.   However, through use of the previously mentioned table of values for the trigonometric functions and a method called "interpolation," we can find approximations for functions of angles not given in the table.   Note, for example, that sin 76°33′ is not shown in Table C-1.   The value of sin 76°33′ lies between the given value for sin 76°30′ and the given value for sin 76°40′.

sin 76°40′ = 0.9730

sin 76°33′ = x

sin 76°30′ = 0.9724

To find the number x, we use linear interpolation. We are assuming that variations between angles are linearly proportional to the variations for values of the functions of the angles. Of course, this should be done only for small intervals and not for intervals in which the angles are close to measures where a function is undefined. That is, using interpolation to find a value such as tan 89°38′ will not result in a high degree of accuracy. The method of interpolation is illustrated by the following examples.

Example 2-6.1   Find sin 49°56′.

Solution

$$10\left[\,7\!\begin{bmatrix}\sin 50°0' = 0.7660 \\ \sin 49°57' = x \\ \sin 49°50' = 0.7642\end{bmatrix}d\,\right]0.0018$$

$$\frac{7}{10} = \frac{d}{0.0018}$$
$$10d = 0.0126$$
$$d = 0.0013$$
$$x = 0.7642 + 0.0013$$
$$x = 0.7655$$

Example 2-6.2   Find cos 49°57′.

Solution

$$10\left[\,7\!\begin{bmatrix}\cos 50°0' = 0.6428 \\ \cos 49°57' = x \\ \cos 49°50' = 0.6450\end{bmatrix}d\,\right]0.0022$$

$$\frac{7}{10} = \frac{d}{0.0022}$$
$$10d = 0.0154$$
$$d = 0.0015$$
$$x = 0.6450 - 0.0015$$
$$x = 0.6435$$

Example 2-6.3   Find θ if tan $\theta = 0.7865$ and $0 < \theta < \dfrac{\pi}{2}$.

Solution

$$10\left[\,d\!\begin{bmatrix}\tan 38°20' = 0.7907 \\ \tan \theta \quad\; = 0.7865 \\ \tan 38°10' = 0.7860\end{bmatrix}0.0005\,\right]0.0047$$

$$\frac{d}{10} = \frac{0.0005}{0.0047} = \frac{5}{47}$$
$$47d = 50$$
$$d = 1$$
$$\theta = 38°10' + 1'$$
$$\theta = 38°11'$$

Example 2·6.4   Find θ if $\cot \theta = 1.782$ and $0 < \theta < \dfrac{\pi}{2}$.

Solution

$$10 \left[d \begin{bmatrix} \cot 29°10' = 1.792 \\ \begin{bmatrix} \cot \theta & = 1.782 \\ \cot 29°20' = 1.780 \end{bmatrix} 0.002 \end{bmatrix} 0.012 \right.$$

$$\frac{d}{10} = \frac{0.002}{0.012} = \frac{2}{12} = \frac{1}{6}$$

$$6d = 10$$

$$d = 2$$

$$\theta = 29°18'$$

Exercise 2·6.1   **1**   Find values for the following:

(a) $\cos 37°17'$ (b) $\tan 62°39'$ (c) $\sec 23°46'$

(d) $\cos 29°2'$ (e) $\csc 37°18'$ (f) $\sin 138°19'$

(g) $\cot 239°5'$ (h) $\sin (-211°)$ (i) $\cos (-137°15')$

(j) $\sec (-475°)$ (k) $\csc 303°11'$ (l) $\tan 255°16'$

2   Find θ if $0 < \theta < \dfrac{\pi}{2}$.

(a) $\cos \theta = 0.8659$ (b) $\sin \theta = 0.8516$ (c) $\tan \theta = 1.387$

(d) $\cot \theta = 1.529$ (e) $\sec \theta = 1.174$ (f) $\csc \theta = 1.369$

(g) $\cos \theta = 0.7193$ (h) $\tan \theta = 7.115$

3 Find θ if $\sin \theta = -0.2250$ and θ is in quadrant III.

4 Find θ if $\tan \theta = -0.6568$ and θ is in quadrant II.

Review Exercise **1** Write the definition of the following trigonometric functions:

    (*a*) $\sin \theta$ (*b*) $\cos \theta$ (*c*) $\tan \theta$

    (*d*) $\csc \theta$ (*e*) $\sec \theta$ (*f*) $\cot \theta$

    (*g*) reciprocal of $\sin \theta$ (*h*) reciprocal of $\cos \theta$

    (*i*) reciprocal of $\tan \theta$

2 Write the values of the six trigonometric functions of an angle θ when a point on the terminal side of the angle is:

    (*a*) (3,4)   (*b*) (−3,4)   (*c*) (−4,−3)

    (*d*) (4,−3)   (*e*) (6,8)   (*f*) (−3,2)

3 Give the reference angle for the following angles in standard position:

    (*a*) 68°   (*b*) 65°45′   (*c*) 100°   (*d*) 120°45′

    (*e*) 200°   (*f*) 240°30′   (*g*) 300°   (*h*) 320°10′

    (*i*) −30°   (*j*) −60°5′   (*k*) −110°   (*l*) −120°10′

    (*m*) 605°   (*n*) 390°   (*o*) 400°   (*p*) 800°

    (*q*) 1320°   (*r*) −1400°   (*s*) 630°   (*t*) 560°

4 Write the values of the following functions without the use of a table:

  (a) sin 30°   (b) csc 60°   (c) cot 60°

  (d) sec 90°   (e) tan 180°   (f) cos 270°

  (g) cot 30°   (h) sec 180°   (i) sin 270°

5 Write the value for each of the following:

  (a) csc 225°   (b) cos 1410°   (c) cot −135°

  (d) cos −300°   (e) sin −600°   (f) csc −390°

  (g) cot −510°   (h) tan −225°   (i) tan 180°

  (j) sec 450°   (k) cot −630°

6 Write the values of the remaining trigonometric functions:

  (a) $\sin \theta = \dfrac{\sqrt{2}}{2}$, $\tan \theta < 0$ (b) $\sin \theta = \dfrac{\sqrt{3}}{2}$, $\tan \theta > 0$

  (c) $\sin \theta = \frac{1}{2}$, $\cot \theta < 0$ (d) $\sin \theta = -\frac{1}{2}$, $\cot \theta > 0$

  (e) $\cot \theta = 1$, $\sin \theta < 0$ (f) $\tan \theta = \sqrt{3}$, $\csc \theta > 0$

  (g) $\sec \theta = -2$, $\tan \theta > 0$

7 Without the use of a table, write the values of the trigonometric functions of the following:

(*a*) cos −270°

(*b*) tan 180°

(*c*) cot 540°

(*d*) sin −450°

(*e*) sec $-\dfrac{3\pi}{2}$

(*f*) csc 6π

8 Write the approximate value of the following trigonometric functions by using a table of values for these functions and interpolating where necessary:

(*a*) sin 8°

(*b*) cos 83°

(*c*) tan 32°

(*d*) csc 67°

(*e*) cot 57°

(*f*) sin 47°

(*g*) cot 20°21′

(*h*) tan 33°40′

(*i*) csc 82°50′

(*j*) sec 67°30′

(*k*) sec 40°10′

(*l*) sin 36°50′

(*m*) sin 21°35′

(*n*) csc 66°25′

(*o*) sec 47°45′

(*p*) cot 38°46′

(*q*) sin 45°48′

(*r*) cos 17°33′

(*s*) tan 73°04′

(*t*) cot 27°48′

(*u*) sin 24°23′

9 Find the angle θ, 0° ≤ θ ≤ 360°, given:

(*a*) sin θ = 0.6381

(*b*) tan θ = 0.6536

(*c*) sec θ = 1.828

(*d*) cot θ = 1.628

(*e*) cos θ = 0.5721

(*f*) csc θ = 1.262

Answers

Exercise 2·1.1 **1** $\sin \theta = \dfrac{t}{r}$ $\cos \theta = \dfrac{s}{r}$

$\tan \theta = \dfrac{t}{s}$ $\cot \theta = \dfrac{s}{t}$

$\sec \theta = \dfrac{r}{s}$ $\csc \theta = \dfrac{r}{s}$

2 (a) $\sin \alpha = \frac{5}{13}$ (b) $\sin \alpha = -\frac{5}{13}$ (c) $\sin \alpha = -\frac{3}{5}$
$\cos \alpha = \frac{12}{13}$ $\cos \alpha = -\frac{12}{13}$ $\cos \alpha = -\frac{4}{5}$
$\tan \alpha = \frac{5}{12}$ $\tan \alpha = \frac{5}{12}$ $\tan \alpha = \frac{3}{4}$
$\cot \alpha = \frac{12}{5}$ $\cot \alpha = \frac{12}{5}$ $\cot \alpha = \frac{4}{3}$
$\sec \alpha = \frac{13}{12}$ $\sec \alpha = -\frac{13}{12}$ $\sec \alpha = -\frac{5}{4}$
$\csc \alpha = \frac{13}{5}$ $\csc \alpha = -\frac{13}{5}$ $\csc \alpha = -\frac{5}{3}$

(d) $\sin \alpha = \frac{1}{2}$ (e) $\sin \alpha = \frac{5}{13}$

$\cos \alpha = \dfrac{\sqrt{3}}{2}$ $\cos \alpha = \pm \frac{12}{13}$

$\tan \alpha = \dfrac{\sqrt{3}}{3}$ $\tan \alpha = \pm \frac{5}{12}$

$\cot \alpha = \sqrt{3}$ $\cot \alpha = \pm \frac{12}{5}$

$\sec \alpha = \dfrac{2\sqrt{3}}{3}$ $\sec \alpha = \pm \frac{13}{12}$

$\csc \alpha = 2$ $\csc \alpha = \frac{13}{5}$

3 (a) $\sin \alpha = -\frac{3}{5}$ (b) $\sin \alpha = -\dfrac{\sqrt{10}}{10}$ (c) $\sin \alpha = -\frac{12}{13}$

$\cos \alpha = \frac{4}{5}$ $\cos \alpha = -\dfrac{3\sqrt{10}}{10}$ $\cos \alpha = \frac{5}{13}$

$\tan \alpha = -\frac{3}{4}$ $\tan \alpha = \frac{1}{3}$ $\tan \alpha = -\frac{12}{5}$

$\cot \alpha = -\frac{4}{3}$ $\sec \alpha = -\dfrac{\sqrt{10}}{3}$ $\cot \alpha = -\frac{5}{12}$

$\sec \alpha = \frac{5}{4}$ $\csc \alpha = -\sqrt{10}$ $\csc \alpha = -\frac{13}{12}$

(d) $\sin \alpha = -\dfrac{\sqrt{10}}{10}$ (e) $\sin \alpha = +\dfrac{\sqrt{10}}{10}$ (f) $\cos \alpha = -\dfrac{\sqrt{3}}{2}$

$\cos \alpha = \dfrac{3\sqrt{10}}{10}$ $\cos \alpha = -\dfrac{3\sqrt{10}}{10}$ $\tan \alpha = \dfrac{\sqrt{3}}{3}$

$\cot \alpha = -3$ $\tan \alpha = -\frac{1}{3}$ $\cot \alpha = \sqrt{3}$

$\sec \alpha = \dfrac{\sqrt{10}}{3}$ $\sec \alpha = -\dfrac{\sqrt{10}}{3}$ $\sec \alpha = -\dfrac{2\sqrt{3}}{3}$

$\csc \alpha = -\sqrt{10}$ $\csc \alpha = \sqrt{10}$ $\csc \alpha = -2$

(g) $\sin \alpha = \dfrac{\sqrt{15}}{4}$

$\tan \alpha = -\sqrt{15}$

$\cot \alpha = -\dfrac{\sqrt{15}}{15}$

$\sec \alpha = -4$

$\csc \alpha = \dfrac{4\sqrt{15}}{15}$

(h) $\sin \alpha = -\dfrac{4\sqrt{41}}{41}$

$\cos \alpha = -\dfrac{5\sqrt{41}}{41}$

$\cot \alpha = \tfrac{5}{4}$

$\sec \alpha = -\dfrac{\sqrt{41}}{5}$

$\csc \alpha = -\dfrac{\sqrt{41}}{4}$

Exercise 2·2.1 **1** (a) 8° (b) 48°43′ (c) 1°19′32″ (d) 39°

(e) 9°38′ (f) 38°43′44″ (g) $\dfrac{\pi}{8}$ (h) $\dfrac{\pi}{5}$

2

θ	$\sin \theta$	$\cos \theta$	$\tan \theta$	$\cot \theta$	$\sec \theta$	$\csc \theta$
120°	$\dfrac{\sqrt{3}}{2}$	$-\dfrac{1}{2}$	$-\sqrt{3}$	$-\dfrac{1}{\sqrt{3}} = -\dfrac{\sqrt{3}}{3}$	-2	$\dfrac{2}{\sqrt{3}} = \dfrac{2\sqrt{3}}{3}$
135°	$\dfrac{\sqrt{2}}{2}$	$-\dfrac{\sqrt{2}}{2}$	-1	-1	$-\sqrt{2}$	$\sqrt{2}$
210°	$-\dfrac{1}{2}$	$-\dfrac{\sqrt{3}}{2}$	$\dfrac{1}{\sqrt{3}} = \dfrac{\sqrt{3}}{3}$	$\sqrt{3}$	$-\dfrac{2}{\sqrt{3}} = -\dfrac{2\sqrt{3}}{3}$	-2
225°	$-\dfrac{\sqrt{2}}{2}$	$-\dfrac{\sqrt{2}}{2}$	1	1	$-\sqrt{2}$	$-\sqrt{2}$
240°	$-\dfrac{\sqrt{3}}{2}$	$-\dfrac{1}{2}$	$\sqrt{3}$	$\dfrac{1}{\sqrt{3}} = \dfrac{\sqrt{3}}{3}$	-2	$-\dfrac{2}{\sqrt{3}} = -\dfrac{2\sqrt{3}}{3}$
300°	$-\dfrac{\sqrt{3}}{2}$	$\dfrac{1}{2}$	$-\sqrt{3}$	$-\dfrac{1}{\sqrt{3}} = -\dfrac{\sqrt{3}}{3}$	2	$-\dfrac{2}{\sqrt{3}} = -\dfrac{2\sqrt{3}}{3}$
330°	$-\dfrac{1}{2}$	$\dfrac{\sqrt{3}}{2}$	$-\dfrac{1}{\sqrt{3}} = -\dfrac{\sqrt{3}}{3}$	$-\sqrt{3}$	$\dfrac{2}{\sqrt{3}} = \dfrac{2\sqrt{3}}{3}$	-2

3 (a) $\sin 480° = \dfrac{\sqrt{3}}{2}$

$\cos 480° = -\tfrac{1}{2}$

$\tan 480° = -\sqrt{3}$

$\cot 480° = -\dfrac{1}{\sqrt{3}} = -\dfrac{\sqrt{3}}{3}$

$\sec 480° = -2$

$\csc 480° = \dfrac{2}{\sqrt{3}} = \dfrac{2\sqrt{3}}{3}$

(b) $\sin \theta = \tfrac{1}{2}$

$\cos \theta = -\dfrac{\sqrt{3}}{2}$

$\tan \theta = -\dfrac{\sqrt{3}}{3}$

$\cot \theta = -\sqrt{3}$

$\sec \theta = -\dfrac{2\sqrt{3}}{3}$

$\csc \theta = 2$

(c) $\sin -\tfrac{11}{6}\pi = \tfrac{1}{2}$

   $\cos -\tfrac{11}{6}\pi = \dfrac{\sqrt{3}}{2}$

   $\tan -\tfrac{11}{6}\pi = \dfrac{1}{\sqrt{3}} = \dfrac{\sqrt{3}}{3}$

   $\cot -\tfrac{11}{6}\pi = \sqrt{3}$

   $\sec -\tfrac{11}{6}\pi = \dfrac{2}{\sqrt{3}} = \dfrac{2\sqrt{3}}{3}$

   $\csc -\tfrac{11}{6}\pi = 2$

(d) $\sin\theta = -\dfrac{\sqrt{3}}{2}$

   $\cos\theta = \tfrac{1}{2}$

   $\tan\theta = -\sqrt{3}$

   $\cot\theta = -\dfrac{1}{\sqrt{3}}$

   $\sec\theta = 2$

   $\csc\theta = -\dfrac{2\sqrt{3}}{3}$

(e) $\sin 585° = -\dfrac{\sqrt{2}}{2}$

   $\tan 585° = 1$
   $\sec 585° = -\sqrt{2}$
   $\cos 585° = -\dfrac{\sqrt{2}}{2}$

   $\cot 585° = 1$

   $\csc 585° = -\sqrt{2}$

(f) $\sin\theta = -\dfrac{\sqrt{3}}{2}$

   $\cos\theta = \tfrac{1}{2}$
   $\tan\theta = -\sqrt{3}$

   $\cot\theta = -\dfrac{1}{\sqrt{3}}$

   $\sec\theta = 2$

   $\csc\theta = -\dfrac{2\sqrt{3}}{3}$

4   (a) $\cos\theta = -\dfrac{\sqrt{3}}{2}$

   $\tan\theta = -\dfrac{1}{\sqrt{3}} = -\dfrac{\sqrt{3}}{3}$

   $\cot\theta = -\sqrt{3}$

   $\sec\theta = -\dfrac{2}{\sqrt{3}} = -\dfrac{2\sqrt{3}}{3}$

   $\csc\theta = 2$

(b) $\sin\theta = \dfrac{3\sqrt{10}}{10}$

   $\cos\theta = -\dfrac{\sqrt{10}}{10}$

   $\cot\theta = -\tfrac{1}{3}$

   $\sec\theta = -\sqrt{10}$

   $\csc\theta = \dfrac{\sqrt{10}}{3}$

(c) $\sin\theta = -\tfrac{1}{2}$

   $\cos\theta = \dfrac{\sqrt{3}}{2}$

   $\tan\theta = -\dfrac{1}{\sqrt{3}} = -\dfrac{\sqrt{3}}{3}$

   $\cot\theta = -\sqrt{3}$
   $\csc\theta = -2$

5 (a) $\dfrac{\sqrt{3}}{2}\left(-\dfrac{\sqrt{3}}{2}\right) - \left(-\dfrac{1}{2}\right)\left(\dfrac{1}{2}\right) = -\dfrac{1}{2}$

$$-\dfrac{3}{4} + \dfrac{1}{4} = -\dfrac{1}{2}$$

(b) $\left(-\dfrac{\sqrt{3}}{2}\right)\left(-\dfrac{1}{2}\right) + \left(\dfrac{1}{2}\right)\left(\dfrac{\sqrt{3}}{2}\right) = \dfrac{\sqrt{3}}{2}$

$$\dfrac{\sqrt{3}}{4} + \dfrac{\sqrt{3}}{4} = \dfrac{\sqrt{3}}{2}$$

(c) $\dfrac{\frac{1}{2} - 1/\sqrt{2}}{2(\frac{1}{2}) + 3(1)} = \dfrac{(1 - \sqrt{2})/2}{4} = \dfrac{1 - \sqrt{2}}{8}$

(d) $1 + (1)^2 = 1 + 1 = 2$

Exercise 2-3.1 **1** (a) $\sin 720° = 0$
$\cos 720° = 1$
$\tan 720° = 0$
$\cot 720° = \text{Undefined}$
$\sec 720° = 1$
$\csc 720° = \text{Undefined}$

(b) $\sin 450° = 1$
$\cos 450° = 0$
$\tan 450° = \text{Undefined}$
$\cot 450° = 0$
$\sec 450° = \text{Undefined}$
$\csc 450° = 1$

(c) $\sin\left(-\dfrac{3\pi}{2}\right) = 1$

$\cos\left(-\dfrac{3\pi}{2}\right) = 0$

$\tan\left(-\dfrac{3\pi}{2}\right) = \text{Undefined}$

$\cot\left(-\dfrac{3\pi}{2}\right) = 0$

$\sec\left(-\dfrac{3\pi}{2}\right) = \text{Undefined}$

$\csc\left(-\dfrac{3\pi}{2}\right) = 1$

(d) $\sin(-\pi) = 0$

$\cos(-\pi) = -1$

$\tan(-\pi) = 0$

$\cot(-\pi) = \text{Undefined}$

$\sec(-\pi) = -1$

$\csc(-\pi) = \text{Undefined}$

(e) $\sin(-7\pi) = 0$
$\cos(-7\pi) = -1$
$\tan(-7\pi) = 0$
$\cot(-7\pi) = \text{Undefined}$
$\sec(-7\pi) = -1$
$\csc(-7\pi) = \text{Undefined}$

(f) $\sin \frac{13}{2}\pi = 1$
$\cos \frac{13}{2}\pi = 0$
$\tan \frac{13}{2}\pi = \text{Undefined}$
$\cot \frac{13}{2}\pi = 0$
$\sec \frac{13}{2}\pi = \text{Undefined}$
$\csc \frac{13}{2}\pi = 1$

(g) $\sin\left(-\dfrac{2\pi}{3}\right) = -\dfrac{\sqrt{3}}{2}$

(h) $\sin 630° = -1$

$\cos\left(-\dfrac{2\pi}{3}\right) = -\tfrac{1}{2}$

$\cos 630° = 0$

$\tan\left(-\dfrac{2\pi}{3}\right) = \sqrt{3}$

$\tan 630° = $ Undefined

$\cot\left(-\dfrac{2\pi}{3}\right) = \dfrac{1}{\sqrt{3}} = \dfrac{\sqrt{3}}{3}$

$\cot 630° = 0$

$\sec\left(-\dfrac{2\pi}{3}\right) = -2$

$\sec 630° = $ Undefined

$\csc\left(-\dfrac{2\pi}{3}\right) = -\dfrac{2}{\sqrt{3}} = -\dfrac{2\sqrt{3}}{3}$

$\csc 630° = -1$

2 (a) $\cos 0° + \sin 270° = 1 + (-1) = 0$

(b) $\sin\dfrac{\pi}{2}\cos\dfrac{3\pi}{2} + \cos\dfrac{\pi}{2}\sin\dfrac{3\pi}{2} = 1\cdot 0 + 0(-1) = 0$

(c) $\cos 2\pi \cos \pi + \sin 2\pi \sin \pi = 1\cdot(-1) + 0\cdot 0 = -1$

(d) $\tan(-\pi)\cot\dfrac{\pi}{2} - \sec 2\pi \cos \pi = 0\cdot 0 - 1(-1) = 1$

Exercise 2-4.1

1 0.9063	**2** 0.85567	**3** 2.414
4 3.7321	**5** 1.942	**6** 1.3980
7 1.561	**8** 1.5597	**9** 0.9217
10 0.99567	**11** 0.0524	**12** 6.3925
13 28.64	**14** 0.15838	**15** Undefined

Exercise 2-5.1

1 −0.3420	**2** −0.9912	**3** −1.540
4 −5.759	**5** −1.618	**6** 0.6494
7 0.1736	**8** Undefined	**9** −0.7716
10 0.2186		

Exercise 2-6.1

1 (a) 0.7957   (b) 1.9333   (c) 1.093

(d) 0.87434   (e) 1.650   (f) 0.66501

(g) 0.5989   (h) 0.51504   (i) −0.7343

(j) −2.3662   (k) −1.195   (l) 3.8028

2 (a) $\theta = 30°1'$ (b) 58°23'   (c) $\theta = 54°13'$ (d) 33°11'

(e) $\theta = 31°37'$ (f) 46°52'   (g) $\theta = 44°$ (h) 82°

3 $\theta = 193°$

4 $\theta = 146°42'$

Review Exercise **1** If $r = \sqrt{x^2 + y^2}$ and (x,y) is a point on the terminal side of θ.

(a) $\dfrac{y}{r}$ (b) $\dfrac{x}{r}$ (c) $\dfrac{y}{x}$ (d) $\dfrac{r}{y}$

(e) $\dfrac{r}{x}$ (f) $\dfrac{x}{y}$ (g) $\dfrac{r}{y}$ (h) $\dfrac{r}{x}$

(i) $\dfrac{x}{y}$

2 (a) $r = 5$, $x = 3$, $y = 4$

$\sin \theta = \tfrac{4}{5}$ $\tan \theta = \tfrac{4}{3}$ $\sec \theta = \tfrac{5}{3}$
$\cos \theta = \tfrac{3}{5}$ $\cot \theta = \tfrac{3}{4}$ $\csc \theta = \tfrac{5}{4}$

(b) $\sin \theta = \tfrac{4}{5}$ $\tan \theta = -\tfrac{3}{4}$ $\sec \theta = -\tfrac{5}{3}$
$\cos \theta = -\tfrac{3}{5}$ $\cot \theta = -\tfrac{4}{3}$ $\csc \theta = \tfrac{5}{4}$

(c) $r = 5$, $x = -4$, $y = -3$

$\sin \theta = -\tfrac{3}{5}$ $\tan \theta = \tfrac{3}{4}$ $\sec \theta = -\tfrac{5}{4}$
$\cos \theta = -\tfrac{4}{5}$ $\cot \theta = \tfrac{4}{3}$ $\csc \theta = -\tfrac{5}{3}$

(d) $\sin \theta = -\tfrac{3}{5}$ $\tan \theta = -\tfrac{3}{4}$ $\sec \theta = \tfrac{5}{4}$
$\cos \theta = \tfrac{4}{5}$ $\cot \theta = -\tfrac{4}{3}$ $\csc \theta = -\tfrac{5}{3}$

(e) $r = 10$, $x = 6$, $y = 8$

$\sin \theta = \tfrac{4}{5}$ $\tan \theta = \tfrac{4}{3}$ $\sec \theta = \tfrac{5}{3}$
$\cos \theta = \tfrac{3}{5}$ $\cot \theta = \tfrac{3}{4}$ $\csc \theta = \tfrac{5}{4}$

(f) $\sin \theta = \dfrac{2}{\sqrt{13}}$ $\tan \theta = -\tfrac{2}{3}$ $\sec \theta = -\dfrac{\sqrt{13}}{3}$

$\cos \theta = -\dfrac{3}{\sqrt{13}}$ $\cot \theta = -\tfrac{3}{2}$ $\csc \theta = \dfrac{\sqrt{13}}{2}$

3 (a) $68°$ (b) $65°45'$ (c) $80°$ (d) $59°15'$
(e) $20°$ (f) $60°30'$ (g) $60°$ (h) $39°50'$
(i) $30°$ (j) $60°5'$ (k) $70°$ (l) $59°50'$
(m) $65°$ (n) $30°$ (o) $40°$ (p) $80°$
(q) $60°$ (r) $40°$ (s) $90°$ (t) $20°$

4 (a) $\dfrac{1}{2}$ (b) $\dfrac{2\sqrt{3}}{3}$ (c) $\dfrac{\sqrt{3}}{3}$

(d) Undefined (e) 0 (f) 0

(g) $\sqrt{3}$ (h) -1 (i) -1

5 (a) $-\sqrt{2}$ (b) $\dfrac{\sqrt{3}}{2}$ (c) 1 (d) $\tfrac{1}{2}$

(e) $\dfrac{\sqrt{3}}{2}$ (f) -2 (g) $\sqrt{3}$ (h) -1

(i) 0 (j) Undefined (k) Undefined

6 (a) $\cos \theta = -\dfrac{\sqrt{2}}{2}$

$\tan \theta = -1$

$\cot \theta = -1$

$\sec \theta = -\sqrt{2}$

$\csc \theta = \sqrt{2}$

(b) $\cos \theta = \frac{1}{2}$

$\tan \theta = \sqrt{3}$

$\cot \theta = \dfrac{\sqrt{3}}{3}$

$\sec \theta = 2$

$\csc \theta = \dfrac{2\sqrt{3}}{3}$

(c) $\cos \theta = -\dfrac{\sqrt{3}}{2}$

$\tan \theta = -\dfrac{\sqrt{3}}{3}$

$\cot \theta = -\sqrt{3}$

$\sec \theta = -\dfrac{2\sqrt{3}}{3}$

$\csc \theta = 2$

(d) $\cos \theta = -\dfrac{\sqrt{3}}{2}$

$\tan \theta = \dfrac{\sqrt{3}}{3}$

$\cot \theta = \sqrt{3}$

$\sec \theta = -\dfrac{2\sqrt{3}}{3}$

$\csc \theta = -2$

(e) $\sin \theta = -\dfrac{\sqrt{2}}{2}$

$\cos \theta = -\dfrac{\sqrt{2}}{2}$

$\tan \theta = 1$

$\sec \theta = -\sqrt{2}$

$\csc \theta = -\sqrt{2}$

(f) $\sin \theta = \dfrac{\sqrt{3}}{2}$

$\cos \theta = \frac{1}{2}$

$\cot \theta = \dfrac{\sqrt{3}}{3}$

$\sec \theta = 2$

$\csc \theta = \dfrac{2\sqrt{3}}{3}$

(g) $\sin \theta = -\dfrac{\sqrt{3}}{2}$

$\cos \theta = -\frac{1}{2}$
$\tan \theta = \sqrt{3}$

$\cot \theta = \dfrac{\sqrt{3}}{3}$

$\csc \theta = -\dfrac{2\sqrt{3}}{3}$

7 (a) 0   (b) 0   (c) Undefined

(d) −1   (e) Undefined   (f) Undefined

8 (a) 0.1392   (b) 0.12187   (c) 0.6249

(d) 1.08637   (e) 0.6494   (f) 0.73135

(g) 2.696   (h) 0.66608   (i) 1.008

(j) 2.61383   (k) 1.309   (l) 0.59949

(m) 0.3678   (n) 1.09113   (o) 1.488

(p) 1.2452   (q) 0.7169   (r) 0.95345

(s) 3.285   (t) 1.8967   (u) 0.4129

9 (a) 39°39′, 140°21′   (b) 33°10′, 213°10′

(c) 56°50′, 303°10′   (d) 31°34′, 211°34′

(e) 55°6′, 304°54′   (f) 52°24′, 127°36′

Solutions to Triangles

3

Objectives

Upon completion of Chapter 3 you will be able to perform the following:

1 **Define the trigonometric functions as ratios of the sides of a right triangle (3-1).**

2 **Solve right triangles (3-2).**

3 **Solve problems involving angles of elevation and depression (3-3).**

4 **Solve oblique triangles (3-4).**

5 **State and use the law of cosines (3-5).**

6 **State and use the law of sines (3-6).**

7 **Given two sides of an oblique triangle and an angle opposite one of the sides, determine if one or more triangles are formed and solve for the remaining parts (3-7).**

3/Solutions to Triangles

3-1 Introduction In Chapter 2 we defined the trigonometric functions of an angle θ in terms of x, y, and r, where $P(x,y)$ is any point on the terminal side of θ and $r = \sqrt{x^2 + y^2}$. We noted that x and y are the sides of a right triangle whose hypotenuse is r, as illustrated in Figure 3-1.1.

Figure 3-1.1

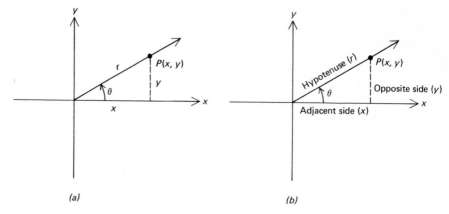

(a)   *(b)*

If θ is an acute angle of a right triangle, the trigonometric functions may be defined as follows:

$$\sin \theta = \frac{\text{length of side opposite } \theta}{\text{length of hypotenuse}}$$

$$\cos \theta = \frac{\text{length of side adjacent to } \theta}{\text{length of hypotenuse}}$$

$$\tan \theta = \frac{\text{length of side opposite } \theta}{\text{length of side adjacent to } \theta}$$

$$\cot \theta = \frac{\text{length of side adjacent to } \theta}{\text{length of side opposite } \theta}$$

$$\sec \theta = \frac{\text{length of hypotenuse}}{\text{length of side adjacent to } \theta}$$

$$\csc \theta = \frac{\text{length of hypotenuse}}{\text{length of side opposite } \theta}$$

3-2 Solving Right Triangles We may now use these definitions to find missing components of a right triangle when the measure of one side and one acute angle or the measure of two sides is known. Finding the missing parts is called "solving the triangle." The following examples illustrate the procedure.

Example 3-2.1   Solve triangle ACB if b is 3, c is 5, and $C = 90°$.

Solution $\cos A = \dfrac{3}{5} = 0.6000$

$A = 53°08'$

$A + B = 90°$

$B = 90° - 53°08'$

$B = 36°52'$

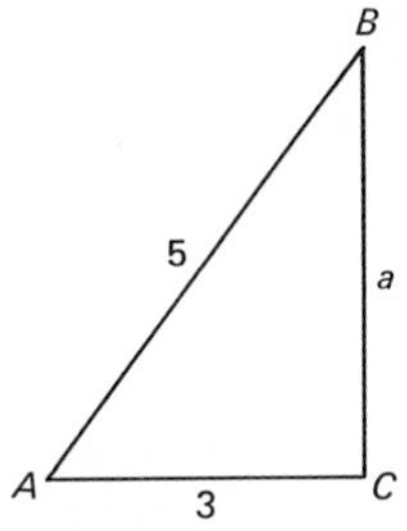

$\sin A = \dfrac{a}{5}$

$\sin 53°08' = \dfrac{a}{5}$

$0.8000 = \dfrac{a}{5}$

$4.000 = a$

The missing components of $\triangle ACB$ are:

     $A = 53°08'$ $B = 36°52'$ $a = 4$

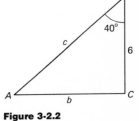

Figure 3-2.1

Example 3-2.2   Solve triangle ACB given $C = 90°$, $B = 40°$, and $a = 6$.

Solution $A = 90° - 40°$ $\tan 40° = \dfrac{b}{6}$

     $= 50°$ $b = 6 \tan 40°$

                     $b = 6(0.8391)$

   $\sec 40° = \dfrac{c}{6}$ $b = 5.035$

   $6 \sec 40° = c$

         $c = 6(1.305)$

         $c = 7.830$

Figure 3-2.2

The missing components of $\triangle ACB$ are

     $A = 50°$ $c = 7.830$ $b = 5.035$

Example 3-2.3   Solve $\triangle ACB$ given $C = 90°$, $A = 47°50'$, and $a = 29.4$.

Solution

$B = 90° - A$

$\quad = 90° - 47°50'$

$\quad = 42°10'$

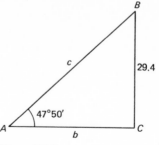

$\cot A = \dfrac{b}{29.4}$

$\quad b = 29.4 \cot 47°50'$

$\quad\quad = (29.4)(0.9057)$ **Figure 3-2.3**

$\quad\quad = 26.63$

$\csc A = \dfrac{c}{29.4}$

$\quad c = 29.4 \csc 47°50'$

$\quad\quad = (29.4)(1.349)$

$\quad\quad = 39.66$

The missing components of $\triangle ACB$ are

$\quad B = 42°10' \quad\quad b = 26.63 \quad\quad c = 39.66$

Exercise 3-2.1   Solve right triangle ACB, given $C = 90°$.

1 $b = 35.0$, $c = 47.5$ **2** $a = 10$, $B = 32°10'$

3 $a = 7.6$, $c = 14.3$ **4** $a = 48.31$, $A = 65°15'$

5 $b = 400$, $a = 150$ **6** $b = 7.281$, $a = 4.165$

7 $a = 23.54$, $B = 87°16'$ **8** $b = 451$, $a = 231$

3-3 Applications of Right Triangles

Solving right triangles has many useful applications. To illustrate, we will show how right triangles are used to find distances and heights that cannot be measured directly.   We will also show how right triangles are used to find the results when a force is applied to an object.   Study the following examples.

Example 3-3.1   A surveyor wishes to know the length of the pond shown in Figure 3-3.1. How can she measure the distance BC?

Solution   She could set up her transit at the point A, 300 ft from C, and sight B, then C. Angle BAC would read 80°.   Now using the tangent of 80°,

$$\tan 80° = \frac{BC}{300 \text{ ft}}$$

$$BC = (300 \text{ ft})(\tan 80°)$$

$$BC = (300 \text{ ft})(5.671)$$

$$BC = 1701.3 \text{ ft}$$

Figure 3-3.1

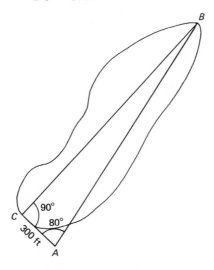

Example 3-3.2   A machinist wants to know the distance between the centers of holes which he must drill, as shown in Figure 3-3.2.

Solution $$\csc 55° = \frac{d}{5 \text{ cm}}$$

$$d = 5 \text{ cm} \times \csc 55°$$

$$= 5 \text{ cm} \times 1.221$$

$$= 6.105 \text{ cm}$$

Figure 3-3.2

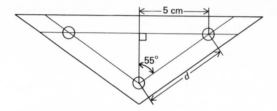

When right triangles are used to find unknown heights or distances from vertical objects, it is customary to refer to angles of elevation and depression. If a point B is above a horizontal line AC, angle BAC is called the *angle of elevation*. If a point B is below a horizontal line AC, angle BAC is called the *angle of depression*. Figure 3-3.3 illustrates these definitions.

Figure 3-3.3

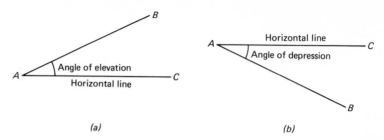

(a)   (b)

The following examples illustrate the use of the angles of elevation and depression:

Example 3-3.3 The angle of elevation of the top of a building as seen from a point 300 ft from the foot of the wall is 47°. How tall is the building?

Solution

$$\tan A = \frac{a}{b}$$

$$\tan 47° = \frac{a}{300}$$

$$a = 300(1.072)$$

$$a = 321.600$$

$$a = 322$$

The height of the building is 322 ft.

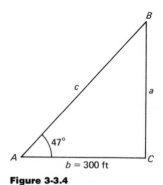

Figure 3-3.4

Example 3-3.4 A man is 40 ft above the ground in an observation tower. He sights a bear some distance away. How far is the bear away from the base of the tower if the line of sight makes an angle of depression of 20° (assume ground is in a horizontal plane)?

Solution   The angle of depression ($\angle MBT$) is 20°; hence,

$$\cot \angle MBT = \frac{d}{40}$$

$$d = 40(\cot 20°)$$

$$= 40(2.747)$$

$$= 109.880$$

The bear is 110 ft away.

Figure 3-3.5

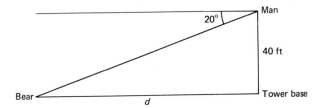

In physics we often need to find the resultant forces when a given force is applied in a particular direction. When a given force is represented graphically on a coordinate system, the effects of this force along the x and y axes are called the "component forces." In Figure 3-3.6, F_x represents the x component of force F, and F_y represents the y component.

Figure 3-3.6

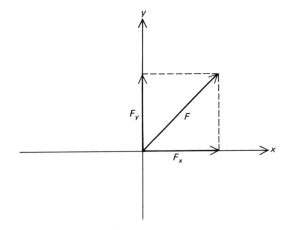

Study the following examples.

Example 3-3.5 A person pushes a sled by applying a 60-lb force downward at 30° below the horizontal. How much of this force is effective in moving the sled across the snow?

Solution The horizontal part of the 60-lb force is its x component.

$$F_x = F \cos \theta$$

$$= (60 \text{ lb})(0.866)$$

$$= 52 \text{ lb}$$

Figure 3-3.7

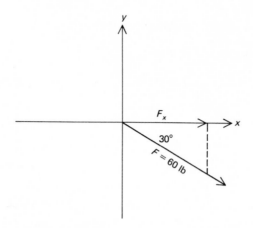

Example 3-3.6 A 140-kg force acts at 23° to the positive x axis. Find the x and y components of this force.

Solution

$$F_x = F \cos \theta \qquad\qquad F_y = F \sin \theta$$

$$= (140 \text{ kg})(\cos 23°) \qquad = (140 \text{ kg})(\sin 23°)$$

$$= (140 \text{ kg})(0.9205) \qquad = (140 \text{ kg})(0.3907)$$

$$= 128.9 \text{ kg} \qquad\qquad = 54.7 \text{ kg}$$

Figure 3-3.8

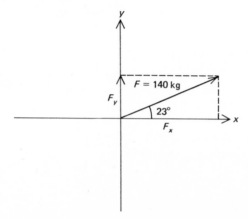

Exercise   3-3.1

1 Points *A* and *C* lie directly opposite each other on the banks of a river. Point *B* is 100 yd downstream from *C*, angle *ACB* measures 90°, and angle *ABC* measures 40°. How wide is the river at *AC*?

2 A person standing at point *A* on a beach sights a boat and a lighthouse at points *B* and *C* respectively. If $\angle ACB = 90°$, $\angle CAB = 50°$, and $\overline{AC}$ is 100 m, how far is the boat from the lighthouse?

3 A woman starts swimming across a river that is 186 m wide. When she reaches the opposite shore, the current has carried her 38 m below the point she wished to reach. How far did she swim?

4 A ship leaves a dock and sails 8 miles out through a channel. It then turns 90° and sails 6 miles. How far is it from the dock?

5 A man standing 150 ft from the foot of a flagpole is 6 ft tall. Find the height of the flagpole if his angle of elevation of its top is 48°.

6 At a horizontal distance of 200 ft from the base of a tower, the angle of elevation of the top is 52°. Find the height of the tower.

7 A person in a balloon observes a house at an angle of depression of 24°. The balloon is 4000 ft high. What is the horizontal distance from the balloon to a point directly over the house?

8 The shadow of a building is 200 ft long. The line from the tip of the shadow to the top of the building makes an angle of inclination of 45°. How high is the building?

9 What is the slope of a line whose angle of elevation is 5°? (Recall that the slope of *A* is the side opposite *A* divided by the side adjacent to *A*.)

10 Find F_x and F_y for:

   (*a*) $F = 40$ lb at an angle of 25° above the positive *x* axis

(b) $F = 100$ g at an angle of 65° above the positive x axis

(c) $F = 30$ lb at an angle of 40° below the positive x axis

11   Find the effective force (F_x) when a person pushes against a car with 85 lb at an angle of 10° below the horizontal (positive x axis).

12   Find the wasted force (F_y) in problem 11 (the force pushing the car into the ground).

13   Find the force lifting a kite if the pull on the string is 10 lb and the kite string makes an angle of 60° with the ground.

3-4   Oblique Triangles

A triangle is called *oblique* if it does not contain a right angle. If one side and two other parts of an oblique triangle are known, the other parts can be found; that is, the triangle can be "solved."

The method used to solve an oblique triangle depends upon which parts are given. There are four possible cases.

CASE I   Given two sides and the included angle.

CASE II   Given three sides.

CASE III   Given one side and two angles.

CASE IV   (The ambiguous case) Given two sides and the measure of an angle opposite one of the sides.

In the following sections we will examine techniques for solving these four cases.

3-5   Law of Cosines

The following statement is known as the

LAW OF COSINES   The square of any side of a triangle is equal to the sum of the squares of the two other sides minus twice the product of these two sides multiplied by the cosine of their included angle.

Using the law of cosines, Figure 3-5.1 shows the three ways in which one side of an oblique triangle may be expressed in terms of the other two sides and the angle opposite side a.

Figure 3-5.1

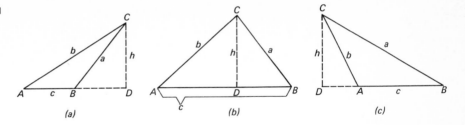

(a)   (b)   (c)

We apply the pythagorean theorem to obtain the following:

In Figure 3-5.1a:

$$a^2 = h^2 + (\overline{BD})^2 = h^2 + (\overline{AD} - c)^2 = h^2 + (\overline{AD})^2 - 2\overline{AD}c + c^2$$

In Figure 3-5.1b:

$$a^2 = h^2 + (\overline{BD})^2 = h^2 + (c - \overline{AD})^2 = h^2 + c^2 - 2\overline{AD}c + (\overline{AD})^2$$

but $h^2 + (\overline{AD})^2 = b^2$ and $\overline{AD} = b \cos A$. Hence, by substitution,

$$a^2 = b^2 + c^2 - 2bc \cos A \qquad \text{(law of cosines)}$$

In Figure 3-5.1c:

$$a^2 = h^2 + (\overline{DA} + c)^2 = h^2 + (\overline{DA})^2 + 2\overline{DA}c + c^2$$

but $h^2 + (\overline{DA})^2 = b^2$ and $\overline{DA} = -b \cos A$. Hence, by substitution,

$$a^2 = b^2 + c^2 - 2bc \cos A \qquad \text{(law of cosines)}$$

Now, since we have designated A and a in all three possible positions in an oblique triangle, it follows that

$$b^2 = a^2 + c^2 - 2ac \cos B \qquad \text{(law of cosines)}$$

and

$$c^2 = a^2 + b^2 - 2ab \cos C \qquad \text{(law of cosines)}$$

The law of cosines may be used to solve Case I (given two sides and an included angle) and Case II (given three sides). The following examples illustrate these cases.

Example 3-5.1   Solve $\triangle ABC$, given $c = 12$, $b = 10$, and $A = 20°$.

Figure 3-5.2

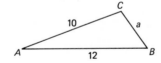

Solution
$$a^2 = b^2 + c^2 - 2bc \cos A$$
$$= 100 + 144 - 2(10)(12)(0.9397)$$
$$= 244 - 225.53$$
$$= 18.47$$
$$a = 4.3$$

$$b^2 = a^2 + c^2 - 2ac \cos B$$
$$100 = 18.47 + 144 - 2(4.3)12 \cos B$$
$$-62.47 = -103.2 \cos B$$
$$\frac{62.47}{103.2} = \cos B$$
$$0.6076 = \cos B$$
$$B = 52°45'$$

$$C = 180° - (A + B)$$
$$= 180° - (20° + 52°45'$$
$$= 180° - 72°45'$$
$$= 107°15'$$

Hence $a = 4.3$, $B = 52°45'$, and $C = 107°15'$.

Example 3-5.2 Solve $\triangle ABC$, given $a = 24$, $b = 18$, and $c = 10$.

Solution
$$a^2 = b^2 + c^2 - 2bc \cos A$$
$$576 = 324 + 100 - 2(18)(10) \cos A$$
$$576 = 424 - 360 \cos A$$
$$\frac{152}{-360} = \cos A$$
$$-0.4222 = \cos A$$
$$114°58' = A$$

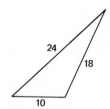

Figure 3-5.3

$$b^2 = a^2 + c^2 - 2ac \cos B$$
$$324 = 576 + 100 - 2(24)(10) \cos B$$
$$-352 = -480 \cos B$$

$$\frac{-352}{-480} = \cos B$$

$$0.7333 = \cos B$$

$$42°50' = B$$

$$C = 180° - (A + B)$$
$$C = 180° - (114°58' + 42°50')$$
$$C = 180° - 157°48'$$
$$C = 22°12'$$

Exercise 3-5.1 Solve $\triangle ABC$ for missing components.

1 $b = 3,\ a = 4,\ C = 52°$ **2** $a = 5,\ b = 8,\ C = 60°$

3 $a = 12,\ c = 15,\ B = 96°$ **4** $b = 20,\ a = 23,\ c = 35$

5 $a = 25,\ b = 20,\ C = 76°50'$ **6** $a = 32,\ c = 40,\ B = 71°45'$

7 $b = 20,\ c = 14,\ A = 110°40'$ **8** $a = 20,\ b = 30,\ c = 25$

9 $a = 125,\ b = 185,\ c = 153$ **10** $a = 40,\ b = 20,\ c = 30$

11 $a = 14,\ b = 8,\ c = 9$ **12** $a = 22.4,\ b = 15.5,\ c = 12.8$

13 $a = 100,\ b = 225,\ c = 75$ **14** $a = 72,\ b = 70,\ c = 80$

15 $a = 47,\ b = 45,\ c = 60$

3-6 Law of Sines

If A, B, and C are the angles of a triangle and a, b, and c are the sides opposite A, B, and C respectively, then

$$\frac{a}{\sin A} = \frac{b}{\sin B} = \frac{c}{\sin C} \qquad \text{(law of sines)}$$

This statement is known as the *law of sines*.

To see why the law of sines is true, we refer to Figure 3-6.1, which shows the three possibilities.

Figure 3-6.1

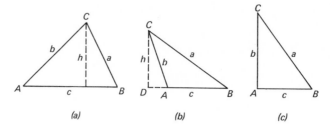

(a)   (b)   (c)

In Figure 3-6.1a, $\sin A = \dfrac{h}{b}$ and $\sin B = \dfrac{h}{a}$. Hence, $b \sin A = h = a \sin B$ and $\dfrac{a}{\sin A} = \dfrac{b}{\sin B}$.

In Figure 3-6.1b, $\sin (180° - A) = \dfrac{h}{b}$ and $\sin B = \dfrac{h}{a}$. Hence, $b \sin (180° - A) = h = a \sin B$, but $\sin (180° - A) = \sin A$; so $b \sin A = a \sin B$ and $\dfrac{a}{\sin A} = \dfrac{b}{\sin B}$.

In Figure 3-6.1c, $\sin A = \sin 90° = 1$, and since $h = b$, $h = b(1) = b \sin A$. But $\dfrac{h}{a} = \sin B$; hence, $b \sin A = h = a \sin B$ and $\dfrac{a}{\sin A} = \dfrac{b}{\sin B}$.

In a similar manner, we can show

$$\frac{a}{\sin A} = \frac{c}{\sin C}$$

hence

$$\frac{a}{\sin A} = \frac{b}{\sin B} = \frac{c}{\sin C}.$$

The law of sines may be used to solve Case III (given one side and two angles). The following examples illustrate this method.

Example 3-6.1   Solve $\triangle ABC$ if $a = 40$, $A = 60°$, and $B = 45°$.

Solution

$$C = 180° - (A + B)$$
$$= 180° - (60° + 45°)$$
$$= 75°$$

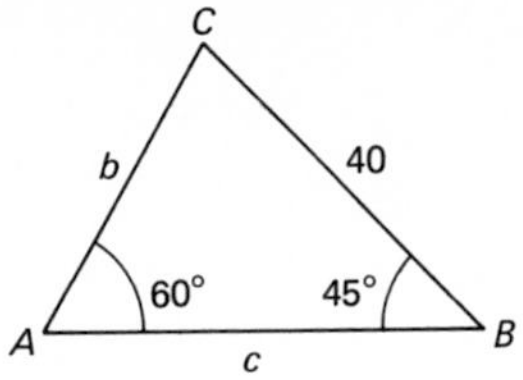

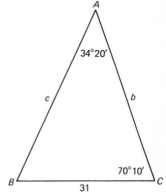

$$\frac{a}{\sin A} = \frac{b}{\sin B}$$

$$40 = a$$

$$\frac{40}{0.8660} = \frac{b}{0.7071}$$

$$b = 32.66$$

Figure 3-6.2

$$\frac{a}{\sin A} = \frac{c}{\sin C}$$

$$\frac{40}{0.8660} = \frac{c}{\sin 75°}$$

$$\frac{40}{0.8660} = \frac{c}{0.9659}$$

$$c = 44.61$$

Hence, $b = 32.66$, $c = 44.61$, and $C = 75°$.

Example 3-6.2   Solve $\triangle ABC$ if $a = 31$, $A = 34°20'$, and $C = 70°10'$.

Solution

$$B = 180° - (A + C)$$
$$= 180° - (34°20' + 70°10')$$
$$= 180° - 104°30'$$
$$= 75°30'$$

$$\frac{a}{\sin A} = \frac{b}{\sin B}$$

$$\frac{31}{0.5640} = \frac{b}{0.9681}$$

$$b = 53.2$$

Figure 3-6.3

$$\frac{a}{\sin A} = \frac{c}{\sin C}$$

$$\frac{31}{0.5640} = \frac{c}{0.9407}$$

$$c = 51.7$$

Hence, $b = 53.2$, $c = 51.7$, and $B = 75°30'$.

Exercise 3-6.1 Solve $\triangle ABC$ given:

1 $a = 28$
$B = 71°$
$C = 46°$

2 $a = 18.4$
$B = 63°$
$C = 54°$

3 $c = 100$
$B = 13°$
$C = 100°$

4 $b = 1.43$
$A = 50°$
$C = 71°$

5 $a = 1500$
$A = 27°16'$
$B = 74°29'$

6 $a = 24.1$
$B = 62°30'$
$C = 42°55'$

3-7 Ambiguous Case When two sides and an angle opposite one of the sides are given, there may be one, two or no triangles determined. This is referred to as the "ambiguous case."

CASE I Suppose that the given angle is right or obtuse; that is, $A \geq 90°$.

(a) If $a \leq b$

We can see by Figure 3-7.1 that no triangle is formed since a is too short.

Figure 3-7.1

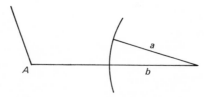

(b) If $a > b$

We can see by Figure 3-7.2 that one triangle can be constructed.

Figure 3-7.2

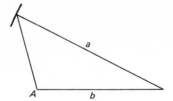

CASE II The given angle is acute; that is, $A < 90°$.

IIA If $a < h$

We can see by Figure 3-7.3 that if $a < h$, a triangle cannot be formed.

Figure 3-7.3

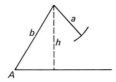

IIB If $a = h$

We can see by Figure 3-7.4 that if $a = h$, one triangle is formed and it is a right triangle.

Figure 3-7.4

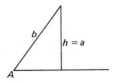

IIC If $h < a < b$

We can see by Figure 3-7.5 that two triangles are formed.

Figure 3-7.5

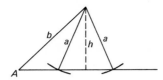

IID If $a \geq b$

We can see by Figure 3-7.6 that one triangle is formed.

Figure 3-7.6

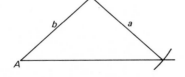

Note the uncertainty that occurs when $A < 90°$ and $a < b$. The decision of whether one, two, or no triangle is determined can be made by applying the law of sines and computing $\sin B$.

Since

$$\frac{\sin B}{b} = \frac{\sin A}{a}$$

it follows that

$$\sin B = \frac{b \sin A}{a} = \frac{h}{a}$$

Now if the computed value for $\sin B$ is greater than 1, a is smaller than h and no triangle is determined; that is

$\sin B > 1$ means $\dfrac{h}{a} > 1$ and $a < h$, and Figure 3-7.3 shows that no triangle can be formed.

If $\sin B = 1$, then $\dfrac{h}{a} = 1$ and $h = a$. Figure 3-7.4 shows that a right triangle is formed.

If $\sin B < 1$, then $\dfrac{h}{a} < 1$ and $h < a$. Since we have assumed $a < b$, we see by Figure 3-7.5 that two triangles are formed.

The following examples illustrate problems involving the ambiguous case.

Example 3-7.1 Solve all triangles for which $B = 108°$, $b = 28$, and $a = 17$.

Solution

$$\frac{\sin A}{17} = \frac{\sin 108°}{28}$$

$$\sin A = \frac{17 \sin 108°}{28}$$

$$= 0.577$$

$$A = 35°$$

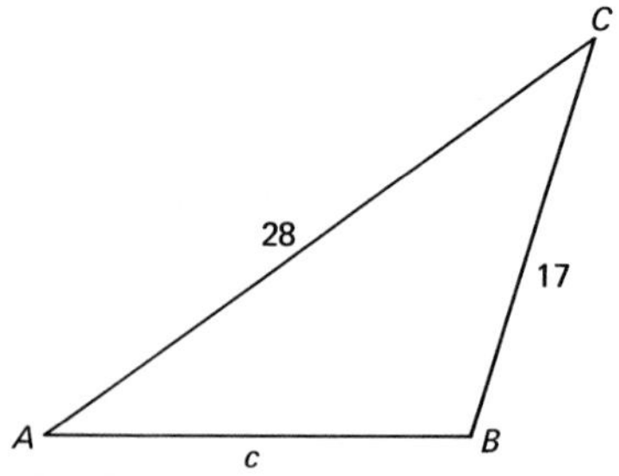

Figure 3-7.7

$B > 90°$ and $a < b$; hence one triangle is formed.

$$C = 180° - (A + B)$$

$$= 180° - 143°$$

$$= 37°$$

$$\frac{c}{\sin C} = \frac{a}{\sin A}$$

$$\frac{c}{0.6018} = \frac{17}{0.5736}$$

$$c = \frac{17(0.6018)}{0.5736}$$

$$c = 17.84$$

Example 3-7.2   Solve all triangles for which $a = 41$, $b = 62$, and $A = 43°$.

Solution

$$\frac{\sin B}{62} = \frac{\sin 43°}{41}$$

$$\sin B = \frac{62 \sin 43°}{41}$$

$$= 1.03$$

Hence no triangle is formed.

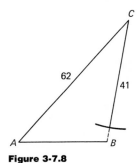

Figure 3-7.8

Example 3-7.3   Solve all triangles for which $a = 12$, $b = 14$, $A = 36°20'$.

Solution

$$\frac{a}{\sin A} = \frac{b}{\sin B}$$

$$\frac{12}{0.5925} = \frac{14}{\sin B}$$

$$\sin B = \frac{4.1475}{6}$$

$$\sin B = 0.6912$$

$$B = 43°43'$$

$$C = 180° - (A + B)$$

$$= 180° - (36°20' + 43°43')$$

$$= 180° - 80°3'$$

$$= 99°57'$$

$$\frac{a}{\sin A} = \frac{c}{\sin C}$$

$$\frac{12}{0.5925} = \frac{c}{\sin 99°57'}$$

$$\frac{12}{0.5925} = \frac{c}{\sin 80°3'}$$

$$\frac{12}{0.5925} = \frac{c}{0.9849}$$

$$c = 19.95$$

Since $A < 90°$, $a < b$, and $a > h$, a second triangle is formed.

$$B' = 180° - B$$
$$= 180° - 43°43'$$
$$= 136°17'$$

$$C' = 180° - (A + B')$$
$$= 180° - (36°20' + 136°17')$$
$$= 180° - 172°37'$$
$$= 7°23'$$

$$\frac{a}{\sin A} = \frac{c'}{\sin C'}$$

$$\frac{12}{0.5925} = \frac{c'}{0.1285}$$

$$c' = 2.6$$

Example 3-7.4   Solve all triangles for which $b = 8.4$, $a = 4.9$, and $B = 21°41'$.

Solution $B < 90°$ and $b > a$; hence there is one triangle.

$$\frac{a}{\sin A} = \frac{b}{\sin B}$$

$$\frac{4.9}{\sin A} = \frac{8.4}{0.3691}$$

$$\sin A = 0.2153$$

$$A = 12°26'$$

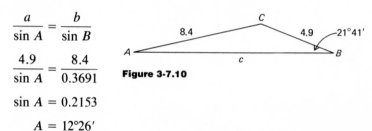

Figure 3-7.10

$$C = 180° - (A + B)$$
$$= 180° - (21°41' + 12°26')$$
$$= 180° - 34°7'$$
$$= 145°53'$$

$$\frac{c}{\sin C} = \frac{b}{\sin B}$$

$$\frac{c}{\sin 145°53'} = \frac{8.4}{0.3691}$$

$$c = 12.76$$

Hence $A = 12°26'$, $C = 145°53'$, and $c = 12.76$.

Exercise 3-7.1 Solve all triangles for which

1 $A = 30°$, $a = 4$, $b = 5$ **2** $A = 20°20'$, $a = 50$, $c = 38$

3 $C = 30°$, $b = 3$, $c = 5$ **4** $B = 65°30'$, $b = 97$, $c = 91$

5 $A = 157°$, $a = 38$, $b = 25$ **6** $C = 148°$, $a = 20$, $c = 26$

7 $A = 76°$, $a = 0.0570$, $b = 0.139$ **8** $B = 71°$, $a = 0.96$, $b = 0.96$

9 $A = 113°$, $a = 12$, $b = 8$ **10** $A = 24°$, $a = 12.6$, $b = 15.7$

11 $A = 40°$, $a = 63$, $c = 42$ **12** $B = 28°$, $a = 13$, $b = 17$

13 $C = 53°50'$, $b = 410$, $c = 379$ **14** $A = 47°10'$, $a = 602$, $b = 807$

15 $A = 5°40'$, $c = 272$, $a = 25$

Review Exercise   **1** Define the trigonometric functions as ratios of the sides of a right triangle.

2 Solve the following right triangles; given $C = 90°$.

(a) $a = 24$, $B = 54°$ (b) $a = 8$, $c = 26$

(c) $a = 120$, $A = 54°$ (d) $a = 12.46$, $c = 17.19$

(e) $A = 6°15'$, $b = 140.9$ (f) $b = 9$, $c = 18$

(g) $a = 15.3$, $B = 23°$ (h) $A = 44°$, $a = 16$

(i) $b = 14$, $a = 10$ (j) $A = 22°$, $b = 10.5$

3 (a) A woman leaves her camp, hikes 4 miles north, then 3 miles east. How far is she from camp?

(b) A man in a boat sights the light of a 100-ft-high lighthouse at an angle of elevation of 3°. How far is the man from the base of the lighthouse?

(c) A person riding in a hot-air balloon sights a landing spot at an angle of depression of 10°. If the balloon is 1000 ft high, how far must the balloonist travel to be directly over the landing spot?

(d) A tugboat pushes against a ship with a force of 6000 lb at an angle of 20°. What amount of this force is pushing the ship forward?

4 Use an appropriate technique to solve the following oblique triangles:

(a) $b = 10$, $A = 70°$, $B = 40°$ (b) $a = 14$, $b = 20$, $A = 38°$

(c) $a = 34.7, b = 20.5, B = 48°$ (d) $a = 15.4, c = 30.7, B = 29°$

(e) $a = 210, b = 300, B = 70°$

5 Use the law of cosines to solve:
   (a) $b = 4, c = 5, A = 60°$ (b) $a = 14, b = 17, c = 20$

   (c) $a = 203, b = 78, C = 62°$ (d) $a = 420, c = 305, B = 88°$

   (e) $a = 12, b = 17, C = 52°$

6 Use the law of sines to solve the following:
   (a) $b = 4, A = 30°, C = 70°$ (b) $a = 18, B = 20°, C = 105°$

   (c) $b = 300, B = 50°, C = 70°$ (d) $a = 50, b = 62, A = 55°$

7 Determine if one or more triangles are formed by the following and solve for the missing information:
   (a) $b = 6, a = 9, c = 8$ (b) $a = 10, c = 15, B = 100°$

   (c) $a = 15, c = 25, A = 38°$ (d) $b = 0.6, a = 0.4, A = 80°$

Answers

Exercise 3-2.1   **1** $A = 42°32'$ **2** $A = 57°50'$
                    $B = 47°28'$ $b = 6.289$
                    $a = 32.1$ $c = 11.813$

3 $A = 32°6'$
   $B = 57°54'$
   $b = 12.11$

4 $B = 24°45'$
   $b = 22.27$
   $c = 53.18$

5 $B = 69°27'$
   $A = 20°33'$
   $c = 427.2$

6 $B = 60°13'$
   $A = 29°47'$
   $c = 8.4$

7 $A = 2°44'$
   $c = 493.63$
   $b = 493.07$

8 $c = 506.7$
   $A = 27°7'$
   $B = 62°53'$

Exercise 3-3.1

1 83.91 yd

2 119.18 m

3 189.84 m

4 10 miles

5 height = 172.59 ft

6 height = 256 ft

7 distance = 8984 ft

8 height = 200 ft

9 slope = 0.0875

10 (a) $F_y = 16.9$ lb
       $F_x = 36.25$ lb

   (b) $F_y = 90.63$ g
       $F_x = 42.26$ g

   (c) $F_y = 19.28$ lb
       $F_x = 22.98$ lb

11 $F_x = 83.71$ lb

12 $F_y = 14.76$ lb

13 $F_y = 8.66$ lb

Exercise 3-5.1

1 $c = 3.2$
   $B = 47°44'$
   $A = 80°16'$

2 $c = 7$
   $A = 38°13'$
   $B = 81°47'$

3 $b = 20.16$
   $C = 47°44'$
   $A = 36°16'$

4 $C = 108°46'$
   $A = 38°28'$
   $B = 32°46'$

5 $c = 28.2$
   $A = 59°34'$
   $B = 43°36'$

6 $b = 42.7$
   $A = 45°23'$
   $C = 62°52'$

7 $a = 28.2$
   $B = 41°38'$
   $C = 27°42'$

8 $A = 41°25'$
   $B = 82°49'$
   $C = 55°46'$

9 $A = 42°6'$
   $B = 82°46'$
   $C = 55°8'$

10 $A = 104°29'$
   $B = 28°56'$
   $C = 46°34'$

11 $A = 110°45'$
   $B = 32°19'$
   $C = 36°56'$

12 $A = 104°15'$
   $B \doteq 42°7'$
   $C = 33°38'$

13 $a^2 = b^2 + c^2 - 2bc \cos A$
   $46{,}250 = 33{,}750 \cos A$
   Not a triangle

14 $A = 56°54'$
   $B = 54°32'$
   $C = 68°34'$

15 $A = 50°45'$
   $B = 47°52'$
   $C = 81°23'$

Exercise 3-6.1	**1** $A = 63°$ $b = 29.7$ $c = 22.6$	**2** $A = 63°$ $b = 18.4$ $c = 16.71$
	3 $A = 67°$ $a = 93.47$ $b = 22.85$	**4** $B = 59°$ $a = 1.28$ $c = 1.58$
	5 $C = 78°15'$ $b = 3154.9$ $c = 3205.6$	**6** $A = 74°35'$ $b = 22.18$ $c = 17.02$

Exercise 3-7.1

1 $B = 141°19'$ or $38°41'$
$C = 8°41'$ or $111°19'$
$c = 1.2$ or 7.45

2 $C = 15°19'$
$B = 144°21'$
$b = 83.87$

3 $B = 17°28'$
$A = 132°32'$
$a = 7.4$

4 $C = 58°36'$
$A = 55°54'$
$a = 88.27$

5 $B = 14°54'$
$C = 8°6'$
$c = 13.7$

6 $A = 24°3'$
$B = 7°57'$
$b = 6.8$

7 No solution

8 $A = 71°$
$C = 38°$
$c = 0.625$

9 $B = 37°51'$
$C = 29°9'$
$c = 6.35$

10 $B = 30°$ or $150°$
$C = 126°$ or $6°$
$c = 25.06$ or 3.24

11 $C = 25°22'$
$B = 114°38'$
$b = 89.09$

12 $A = 21°02'$
$C = 130°58'$
$c = 27.34$

13 $B = 60°50'$ or $119°10'$
$A = 65°20'$ or $7°$
$a = 426.65$ or 57.21

14 $B = 79°25'$ or $100°35'$
$C = 53°25'$ or $32°15'$
$c = 659.2$ or 440

15 No solution

Review Exercise

1 $\sin x = \dfrac{\text{opposite}}{\text{hypotenuse}}$ $\cos x = \dfrac{\text{adjacent}}{\text{hypotenuse}}$

$\tan x = \dfrac{\text{opposite}}{\text{adjacent}}$ $\cot x = \dfrac{\text{adjacent}}{\text{opposite}}$

$\sec x = \dfrac{\text{hypotenuse}}{\text{adjacent}}$ $\csc x = \dfrac{\text{hypotenuse}}{\text{opposite}}$

2 (a) $A = 36°$
$b = 33.04$
$c = 40.83$

(b) $A = 17°55'$
$B = 72°05'$
$b = 24.75$

(c) $B = 36°$
    $c = 148.33$
    $b = 87.2$

(d) $A = 46°24'$
    $B = 43°36'$
    $b = 11.85$

(e) $c = 141.75$
    $a = 15.4$
    $B = 83°45'$

(f) $B = 30°$
    $A = 60°$
    $a = 9\sqrt{3}$

(g) $A = 67°$
    $c = 16.62$
    $b = 6.49$

(h) $B = 46°$
    $c = 23.04$
    $b = 16.58$

(i) $A = 35°32'$
    $B = 54°28'$
    $c = 17.2$

(j) $B = 68°$
    $a = 4.24$
    $c = 11.32$

3 (a) 5 miles

(b) $d = 1908$ ft

(c) $d = 5671$ ft

(d) 5638 lb

4 (a) $C = 70°$
    $a = 14.62$
    $c = 14.62$

(b) $B = 61°36'$
    $C = 80°24'$
    $c = 22.42$

(c) No triangle

(d) $C = 127°34'$
    $A = 23°26'$
    $b = 18.77$

(e) $A = 41°08'$
    $C = 68°52'$
    $c = 297.76$

5 (a) $C = 70°53'$
    $B = 49°07'$
    $a = \sqrt{21}$

(b) $A = 43°32'$
    $B = 56°45'$
    $C = 79°43'$

(c) $A = 95°32'$
    $B = 22°28'$
    $c = 180.15$

(d) $A = 55°20'$
    $C = 36°40'$
    $b = 510.35$

(e) $A = 44°31'$
    $B = 83°29'$
    $c = 13.48$

6 (a) $B = 80°$
    $a = 2.03$
    $c = 3.82$

(b) $A = 55°$
    $b = 7.52$
    $c = 21.22$

(c) $A = 60°$
    $a = 339.16$
    $c = 368.03$

(d) No triangle

7 (a) $A = 78°35'$
    $B = 40°41'$
    $C = 60°44'$

(b) $A = 30°28'$
    $b = 19.4$
    $C = 49°32'$

(c) No triangle

(d) No solution

Application of Radian Measure

Objectives

Upon completion of Chapter 4 you will be able to perform the following:

1 Find the length of the arc intercepted by a central angle θ in a circle of radius r. (4-2)

2 Given the radius r of a circle, determine the linear and angular velocity of a point traveling along the circumference of the circle. (4-3)

3 Given the radius r and a central angle θ, find the area of the circular sector generated by θ. (4-4)

4 Given the radius r and a central angle AOB, find the area of the circular segment AB. (4-4)

4/Application of Radian Measure

4-1   Introduction   In Chapter 1 we introduced a unit of angular measure called the radian. We saw that π radians = 180 degrees or 1 radian = $\dfrac{180}{\pi}$ degrees. Radian measure is very useful because if we use the radian for the unit of measure, we may treat it as a real number. Example 4-1.1 illustrates this idea.

Example 4-1.1   Find the length of a degree on the equator if the diameter of the earth is taken as 7912 miles.

Solution   Using the formula for arc length that we will introduce in the next section, we see

$$l = r\theta$$

$$l = 3956 \text{ miles} \times 1 \text{ degree}$$

$$l = 3956 \text{ miles} \times \frac{\pi}{180} \text{ radians}$$

Now, considering a radian as a real number, we have

$$\frac{\pi}{180} \text{ radians} = \frac{\pi}{180}$$

Hence,

$$l = 3956 \text{ miles} \times \frac{\pi}{180}$$

$$l = \frac{3956 \times \pi}{180} \text{ miles}$$

In this chapter we will see how radian measure is used to calculate certain arc lengths, areas of sectors and segments of circles, and linear and angular velocity.

4-2   Length of an Arc of a Circle   Calculating the true length of an arc of a curve is difficult and requires concepts of the calculus. We can, however, use trigonometry to approximate the length of an arc of a circle. Consider the circle with radius r shown in Figure 4-2.1.

Figure 4-2.1

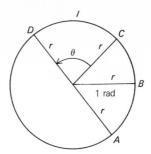

Let θ be the central angle that intercepts arc CD. If θ is measured in radians, then $\dfrac{l}{\theta} = \dfrac{r}{1}$ because in the same circle, or in congruent circles, central angles are proportional to their intercepted arcs. This yields the formula

$$l = r\theta$$

where r is the radius of the circle and l is the length of an arc intercepted by the central angle θ.

The following examples demonstrate the use of this formula.

Example 4-2.1   In a circle with radius 14, find the length of the arc intercepted by the central angle with measure $\dfrac{\pi}{2}$.

Solution
$$l = r\theta$$
$$= 14\left(\frac{\pi}{2}\right)$$
$$= 7\pi$$

Example 4-2.2   Find the number of degrees in the central angle of a circle with a radius of 40 ft if the angle subtends an arc of 26 ft.

Solution
$$l = r\theta$$
$$26 = 40\theta$$
$$\theta = \tfrac{26}{40}\text{ radian}$$
$$= \tfrac{13}{20}\text{ radian}$$
$$= \frac{13}{20} \cdot \frac{180}{\pi}\text{ degrees}$$

$$= \frac{13 \cdot 9}{\pi} \text{ degrees}$$

$$= 37.2 \text{ degrees}$$

Exercise 4-2.1   Find the missing component in the formula $l = r\theta$ for problems 1–6.

1 l when $r = 18$ and $\theta = \frac{3}{4}\pi$

2 l when $r = 6$ and $\theta = \frac{1}{2}\pi$

3 r when $l = 27$ and $\theta = \pi$

4 r when $l = 16.4$ and $\theta = \frac{2}{3}\pi$

5 θ when $r = 20$ and $l = 47$

6 θ when $r = 8.3$ and $l = 40$

7 The end of a 35-in pendulum swings through a 4-in arc. What is the size of the angle through which the pendulum swings?

8 Find the length of the tread mark made by a wheel turning one revolution if its radius is 20 in.

9 If the radius of a sphere is 20 in, what is the length of an arc intercepted by a central angle of 60°?

4-3 Linear and Angular Velocity   If an object moves with constant velocity v, through a distance s in time t, then $s = vt$ or $v = \dfrac{s}{t}$. The *linear velocity* of the object is called v. The formula holds true regardless of the path of the moving object.

Now consider a point P which moves along the circumference of a circle from point A to point B in time t, as shown in Figure 4-3.1. The distance traveled in time t is $s = AB$, and the central angle generated is θ. Now $\dfrac{\theta}{t}$ is called the *angular velocity* of the point P. If θ is measured in radians, $AB = r\theta$, as we have seen in Section 4-2. Hence, $\dfrac{s}{t} = \dfrac{AB}{t} = \dfrac{r\theta}{t}$. Now if we let $\dfrac{\theta}{t} = \omega$, we see $v = r\omega$, or *the linear velocity is equal to the radius times the angular velocity.*

The following examples illustrate solutions to problems involving linear and angular velocity.

Figure 4-3.1

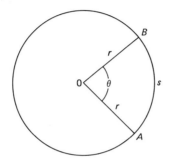

Example 4-3.1 The angular velocity of a flywheel is 1000 radians/min. Find the linear velocity of a point on the circumference of the flywheel if its diameter is 2 m.

Solution Since $v = r\omega$, $r = 1$ m, and $\omega = 1000$ radians/min,

$$v = 1 \cdot 1000 \text{ m/min}$$

$$= 1000 \text{ m/min}$$

Example 4-3.2 The propeller of an airplane has a 4-ft radius, the linear velocity of the tip of the propeller is 200 ft/s. What is the angular velocity?

Solution $\quad v = r\omega \qquad\qquad \omega = \dfrac{v}{r}$

$\qquad\qquad v = 200 \text{ ft/s}$

$\qquad\qquad\qquad\qquad\qquad = \frac{200}{4}$

$\qquad\qquad r = 4 \text{ ft}$

$\qquad\qquad\qquad\qquad\qquad = 50 \text{ radians/s}$

Exercise 4-3.1 **1** If the angular velocity of a wheel is 270 radians/min, what is its linear velocity in terms of r?

2 What is the linear velocity of a point on the circumference of the wheel in problem 1 if it has a radius of 2 ft?

3 The angular velocity of a flywheel is 400 radians/min. Find the linear velocity of a point halfway out on the radius if the wheel has a diameter of 2 ft.

4 A bicycle wheel has a diameter of 28 in. How many revolutions does it make in 1 min if a point on the tread of the tire has a linear velocity of 56π in/min?

5 If the angular velocity of a bicycle wheel with radius 28 in is 8 radians/s, how long will it take the bike to travel 336 in?

6 The minute hand of a clock is 2 cm long. What is the linear velocity of the tip of the hand?

4-4 Areas of Sectors and Segments of Circles

A *sector* of a circle is the figure bounded by two radii and the arc intercepted by them. Figure 4-4.1 shows the sector OAB. From plane geometry we know that the area of a sector of a circle is equal to one-half the product of the radius and the arc. Hence

Area $OAB = \frac{1}{2}rAB$

but

$AB = r\theta$

Thus

Area $OAB = \frac{1}{2}r(r\theta)$

$= \frac{1}{2}r^2\theta$

Figure 4-4.1

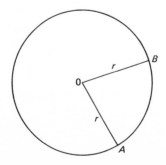

Now if k represents the area of a sector of a circle with radius r, $k = \frac{1}{2}r^2\theta$, where θ is expressed in radians.

A *segment* of a circle is the figure bounded by a chord and its arc. Figure 4-4.2 illustrates the segment of a circle. The area of segment AB is equal to the area of the sector OAB minus the area of the triangle OAB.

Now the area of $\triangle OAB$ may be shown to be

$$\text{Area } \triangle OAB = \tfrac{1}{2}r^2 \sin \theta$$

and the area of the sector OAB is $\frac{1}{2}r^2\theta$. Hence the area of the segment is

$$\text{Area of segment } AB = \tfrac{1}{2}r^2\theta - \tfrac{1}{2}r^2 \sin \theta$$

$$= \tfrac{1}{2}r^2(\theta - \sin \theta)$$

Figure 4-4.2

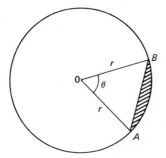

The following examples illustrate problems involving areas of sectors and segments.

Example 4-4.1   Find the area of a sector of a circle whose central angle is $\dfrac{2\pi}{3}$ and whose diameter is 30 cm.

Solution

$$k = \tfrac{1}{2}r^2\theta$$

$$= \tfrac{1}{2}(15)^2 \tfrac{2}{3}\pi$$

$$= 75\pi \text{ cm}^2$$

Example 4-4.2   Find the area of a segment of a circle with diameter 24 in if its central angle is 30°.

Solution

$$k = \tfrac{1}{2}r^2(\theta - \sin \theta)$$

$$= \tfrac{1}{2}(12)^2\left(\frac{\pi}{6} - \frac{1}{2}\right)$$

$$= \tfrac{1}{2}(144)\left(\frac{\pi}{6} - \frac{3}{6}\right)$$

$$= 72\left(\frac{\pi - 3}{6}\right)$$

$$= (12\pi - 36) \text{ in}^2$$

NOTE Since θ is in radians, you may need to convert from radians to degrees by the methods of Chapter 2 in order to find $\sin \theta$ in the formula for the area of a segment of a circle.

Exercise 4-4.1 Find the areas of the sector and the segment of a circle for the given values of the radius r and central angle θ.

1 $r = 5$ cm, $\theta = \dfrac{\pi}{3}$ **2** $r = 18$ in, $\theta = \dfrac{\pi}{6}$

3 $r = 2$ ft, $\theta = \dfrac{5\pi}{12}$ **4** $r = 10$ in, $\theta = \dfrac{11\pi}{6}$

5 $r = 15$ m, $\theta = \dfrac{\pi}{4}$

Review Exercise Calculate the length of the arc, the area of the sector, and the area of the segment of the circle for the given values of the radius r and the central angle θ in problems 1–3.

1 $r = 20$ in, $\theta = \dfrac{\pi}{6}$ **2** $r = 32$ cm, $\theta = \dfrac{3\pi}{4}$

3 $r = 16$ in, $\theta = \dfrac{2\pi}{3}$

4 An automobile traveling at a speed of 10 mph has wheels 30 in in diameter. Find the angular velocity in revolutions per second. **Hint:** 60 mph = 88 ft/s.

5 Find the linear velocity of the tip of the minute hand of a clock if the hand is 6 in long.

6 Two pulleys are connected by a belt. The small pulley has a diameter of 6 cm. If the radius of the larger pulley is 7 cm, find the ratio of the angular speed of the small pulley to the larger one when the belt is in motion. **Hint:** The linear speed of the belt must be equal on each pulley.

7 A satellite orbits the earth at 20,000 mph in a circular path of radius 4000 miles. Find its angular velocity in radians per hour.

8 A bicycle is traveling 10 mph (14.67 ft/s). Its wheels are 28 in in diameter. What is the angular velocity of the wheels in radians per second?

Answers

Exercise 4-2.1

1 $l = \frac{27}{2} \pi$

2 $l = 3\pi$

3 $r = \dfrac{27}{\pi}$

4 $r = \dfrac{24.6}{\pi}$

5 $\theta = 2.35$

6 $\theta = 4.82$

7 0.1142 radian

8 40π in

9 $\dfrac{20\pi}{3}$ in

Exercise 4-3.1

1 $270r$

2 540 ft/min

3 200 ft/min

4 2 rev/min

5 $\frac{3}{2}$ s

6 $\dfrac{\pi}{15}$ cm/min

Exercise 4-4.1

1 $\dfrac{25\pi}{6}$ cm², $\dfrac{25}{2}\left(\dfrac{2\pi - 3\sqrt{3}}{6}\right)$ cm²

2 27π in², $162\left(\dfrac{\pi - 3}{6}\right)$ in²

3 $\dfrac{5\pi}{6}$ ft², 0.686 ft²

4 $\frac{275}{3}\pi$ in², $\frac{25}{3}(11\pi + 3)$ in²

5 $\dfrac{225\pi}{8}$ m², $\dfrac{225}{8}(\pi - 2\sqrt{2})$ m²

Review Exercise

1 $l = \frac{10}{3} \pi$ in, $k = \frac{100}{3} \pi$ in², $k = 200\left(\dfrac{\pi - 3}{6}\right)$ in²

2 $l = 24\pi$ cm, $k = 384\pi$ cm², $k = 128(3\pi - 2\sqrt{2})$ cm²

3 $l = \frac{32}{3} \pi$ in, $k = \frac{256}{3} \pi$ in², $k = 64\dfrac{4\pi - 3\sqrt{3}}{3}$ in²

4 11.7 rad/s

5 $\dfrac{\pi}{5}$ in

6 $\frac{7}{3}$

7 5 rad/h

8 12.6 rad/s

Trigonometric Functions of Real Numbers

5

5/Trigonometric Functions of Real Numbers

5-1 Introduction In Chapter 3 the trigonometric functions were shown as functions from the set of angle measures to the set of real numbers. The domain of the functions was a subset of the set of degrees or a subset of the set of radians, and the range was a subset of the real numbers. Thus, in effect, each trigonometric function associated an angle with a real number.

Since radians may be considered as real numbers, each function had a subset of the real numbers as its domain and a subset of the real numbers as its range. This definition of the trigonometric functions was, however, based upon angle measure. In this chapter we will again show the trigonometric functions as functions with a subset of the real numbers as their domains and a subset of the real numbers as their ranges, but *we will not use angles in our definitions*.

You may be wondering why it is important to define the trigonometric functions of real numbers without the use of angles. The answer is that while angles are fine for surveying, drafting, navigation, etc., scientists and mathematicians are often involved in mathematical analyses that require only real numbers. It becomes important in these instances that they be able to apply the trigonometric functions directly to the real numbers without reference to angles.

5-2 Expansion Formulas (Optional) The following formulas for the sine and cosine of a real number are called the "expansion formulas." Although the proofs of these formulas lie beyond the scope of this book, you may see them later in your study of the calculus.

If t is a real number, then

$$\sin t = t - \frac{t^3}{3!} + \frac{t^5}{5!} - \frac{t^7}{7!} + \cdots$$

and

$$\cos t = 1 - \frac{t^2}{2!} + \frac{t^4}{4!} - \frac{t^6}{6!} + \cdots$$

[Recall that $n! = n(n - 1)(n - 2) \cdots (3)(2)(1)$]

Using the above definitions, we can define the remaining trigonometric functions as follows:

$$\tan t = \frac{\sin t}{\cos t} \qquad \cot t = \frac{1}{\tan t}$$

$$\sec t = \frac{1}{\cos t} \qquad \csc t = \frac{1}{\sin t}$$

Example 5-2.1 Use the expansion formula to find sin 1.

Solution $$\sin 1 = 1 - \frac{1^3}{3!} + \frac{1^5}{5!} - \frac{1^7}{7!}$$

$$= 1 - \frac{1}{(3)(2)} + \frac{1}{(5)(4)(3)(2)} - \frac{1}{(7)(6)(5)(4)(3)(2)}$$

$$= 1 - \frac{(7)(6)(5)(4)}{(7)(6)(5)(4)(3)(2)} + \frac{(7)(6)}{(7)(6)(5))4)(3)(2)} - \frac{1}{(7)(6)(5)(4)(3)(2)}$$

$$= 1 - 0.166667 + 0.008333 - 0.000198$$

$$= 0.8415$$

The expansion formulas are actually used to construct all tables of values of the trigonometric functions, including the tables of values for degrees. (Can you explain how this can be done?)

Table C-2 in the back of the book gives the values of all the trigonometric functions for real numbers between 0 and 1.57. You should use the expansion formulas to verify several of these entries.

Exercise 5-2.1 (Optional) Use the expansion formulas to verify the following (you may want to use a calculator):

1 sin 0.30 = 0.2955   **2** cos 0.80 = 0.6967   **3** tan 1.20 = 2.572

5-3 Circular Functions In Section 5-2 we defined the trigonometric functions of real numbers in terms of the expansion formulas. Since the proofs of the expansion formulas require a knowledge of the calculus, we now turn to definitions more suited to our study of trigonometry. We will define the trigonometric functions of the real numbers in terms of coordinates of points on the "unit circle." The "unit circle" is the graph of the equation $x^2 + y^2 = 1$. When defined in this way, the trigonometric functions are usually referred to as "the circular functions."

To set the stage for our definitions, we proceed as illustrated in Figures 5-3.1 and 5-3.2.

Figure 5-3.1

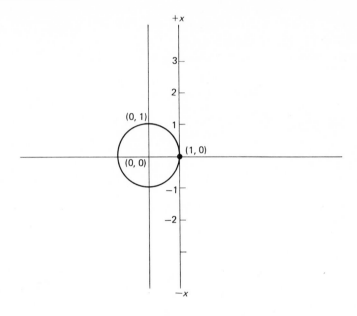

Figure 5-3.2

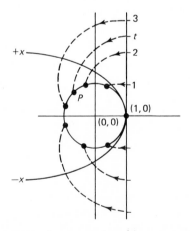

Through the point (1,0) on the unit circle we construct a real number line in a vertical position with unit length the same as the units of the coordinate system. We now visualize the winding of this number line about the circumference of the unit circle. (See Figure 5-3.2.) The upper ray which represents the positive real numbers is wound in a counterclockwise direction, and the lower ray which represents the negative real numbers is wound in a clockwise direction. Hence, *every* real number is associated with a unique point on the graph of the unit circle. We note, however, that although each number is represented by a unique point, each point represents infinitely many real numbers. If a point P represents a real number t, it also represents every real number $t + 2k\pi$ where k is an integer (with each revolution another number will fall on P), that is, . . . , $t - 4\pi$, $t - 2\pi$, t, $t + 2\pi$, $t + 4\pi$, ...

We now define the circular functions as follows:

If t is a real number associated with the point $P(x,y)$ on the unit circle, then

$$\sin t = y \qquad\qquad \cos t = x$$

$$\tan t = \frac{y}{x} \quad x \ne 0 \qquad \cot t = \frac{x}{y} \quad y \ne 0$$

$$\sec t = \frac{1}{x} \quad x \ne 0 \qquad \csc t = \frac{1}{y} \quad y \ne 0$$

Figure 5-3.3 illustrates these concepts.

Figure 5-3.3

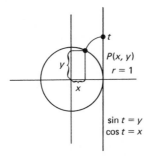

Since x values are positive for points that lie in quadrants I and IV and negative in quadrants II and III, the cosine values for real numbers that correspond to points in quadrants I and IV will be positive, while cosine values of real numbers corresponding to points in quadrants II and III will be negative. In a similar way, $\sin t$ will be positive if t corresponds to a point in quadrant I or II and negative when t corresponds to a point in quadrant III or IV. See Figure 5-3.4.

Figure 5-3.4

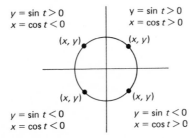

Example 5-3.1   Write the circular functions for t if t corresponds to the point $P\left(-\dfrac{\sqrt{3}}{2}, \dfrac{1}{2}\right)$.

Solution

$$\sin t = \frac{1}{2} \qquad\qquad\qquad \cos t = -\frac{\sqrt{3}}{2}$$

$$\tan t = -\frac{1}{\sqrt{3}} = \frac{-\sqrt{3}}{3} \qquad \cot t = -\sqrt{3}$$

$$\sec t = -\frac{2}{\sqrt{3}} = \frac{-2\sqrt{3}}{3} \qquad \csc t = 2$$

Exercise 5-3.1   Write the circular functions for t if t is a real number that corresponds to the following points on the unit circle:

1 $P\left(\dfrac{1}{\sqrt{2}}, \dfrac{1}{\sqrt{2}}\right)$ **2** $P(1,0)$ **3** $P\left(-\dfrac{1}{\sqrt{2}}, \dfrac{1}{\sqrt{2}}\right)$

4 $P(-1,0)$ **5** $P\left(\dfrac{\sqrt{3}}{2}, \dfrac{1}{2}\right)$ **6** $P\left(-\dfrac{1}{2}, -\dfrac{\sqrt{3}}{2}\right)$

5-4 Circular Functions of Special Numbers   We note that along the unit circle there are points with coordinates which are easily remembered. These points represent the real numbers $0, \dfrac{\pi}{6}$, $\dfrac{\pi}{4}, \dfrac{\pi}{3}, \dfrac{\pi}{2}$, and multiples of these numbers, as illustrated by Figure 5-4.1.

The real numbers $0, \dfrac{\pi}{2}, \pi, \dfrac{3\pi}{2}$, and 2π correspond to points on the co-ordinate axes and may be called "quadrantal points." The sine and cosine values for these numbers are ± 1 and 0. Hence $\sin \dfrac{\pi}{2} = 1$, $\cos \dfrac{3\pi}{2} = 0$, etc.

Note the relationship between the functional values of these numbers.

$$\sin \frac{5\pi}{6} = \frac{1}{2} = \sin \frac{\pi}{6}$$

$$\cos \frac{7\pi}{6} = -\frac{\sqrt{3}}{2} = \cos \frac{5\pi}{6}$$

Figure 5-4.1

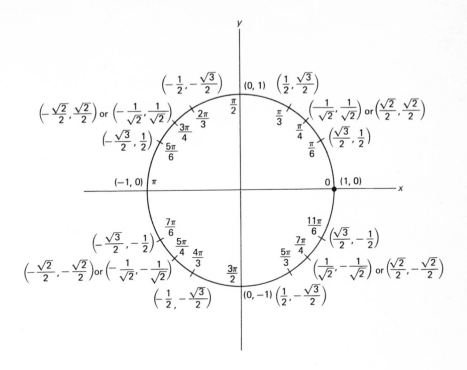

You should refer to Figure 5-4.1 and find other relationships.
The technique illustrated in Section 2-2 may be used to help you recall the functional values of the special numbers.

Exercise 5-4.1   Without the use of a table evaluate the following:

1 $\cos \dfrac{\pi}{3}$ **2** $\tan \dfrac{\pi}{2}$ **3** $\cot \dfrac{3\pi}{4}$

4 $\sec \dfrac{11\pi}{6}$ **5** $\csc \dfrac{4\pi}{3}$ **6** $\sin \dfrac{7\pi}{6}$

Give a second real number that has the same value as:

7 $\sin \pi$ **8** $\cos \dfrac{5\pi}{3}$ **9** $\tan \dfrac{2\pi}{3}$

10 $\csc \dfrac{4\pi}{3}$ **11** $\sec \dfrac{7\pi}{4}$ **12** $\cot \dfrac{5\pi}{4}$

5-5 Evaluating Circular Functions with a Table

A complete table of the values of the circular functions for any real number t, where $0 \le t \le 1.57$, is given in Table C-2.

To find the value of a circular function of a real number t when $0 \le t \le 1.57$, we look up the value directly, interpolating when necessary.

Example 5-5.1 Find $\sin 0.93$.

Solution $\sin 0.93 = 0.8016$ (from Table C-2)

Example 5-5.2 Find $\cot 0.957$.

Solution

$$0.010 \left[0.007 \begin{bmatrix} \cot 0.950 = 0.7151 \\ \cot 0.957 = x \end{bmatrix} d \atop \cot 0.960 = 0.7001 \right] 0.0150$$

Hence,

$$\frac{0.007}{0.010} = \frac{d}{0.0150}$$

$$\frac{7}{10} = \frac{d}{0.0150}$$

$$10d = 0.1050$$

$$d = 0.0105$$

$$\begin{array}{r} 0.7151 \\ -\ 0.0105 \\ \hline 0.7046 \end{array}$$

$$\cot 0.957 = 0.7046$$

To find the circular functional values for a number t greater than 1.57, we must proceed in three steps:

STEP 1 Find a number t_1 such that t_1 is less than 2π and is represented by the same point on the unit circle.

STEP 2 Find a number t_r such that t_r is less than 1.57 and t_r has the same numerical coordinates as t_1 except, perhaps, for the sign.

STEP 3 Look up the values of the circular function of t_r; then assign the proper sign.

This process is very much like the use of reference angles which we described in Chapter 2. Indeed, the number t_r is actually a "reference" number, hence the subscript r.

We noted in Section 5-3 that if a point $P(x,y)$ on the unit circle represents a real number t, it also represents the real number $t + 2k\pi$ where k is an integer. In Step 1 we exchange (if necessary) our real number t for a number t_1 which is represented by the same point as t. Hence, if $t = t_1 + 2k\pi$, we use t_1 instead of t. Now $0 \le t_1 \le 2\pi$, and we are ready to find a "reference" number t_r such that $0 \le t_r \le 1.57$ that has the same circular functional values as t_1 except for sign. To find the proper values of t_r, we may refer to the following:

$$\text{In quadrant I} \qquad t_r = t_1$$

$$\text{In quadrant II} \qquad t_r = \pi - t_1 \quad \text{or} \quad 3.1416 - t_1$$

$$\text{In quadrant III} \qquad t_r = t_1 - \pi \quad \text{or} \quad t_1 - 3.1416$$

$$\text{In quadrant IV} \qquad t_r = 2\pi - t_1 \quad \text{or} \quad 6.280 - t_1$$

The following examples illustrate this method (use $\pi \approx 3.1416$).

Example 5-5.3 Find $\sin 2.800$.

Solution $\qquad t = 2.800$

hence

$$t_1 = 2.800$$

t_1 is represented by a point in quadrant II, hence

$$t_r = 3.1416 - t_1$$

$$= 3.1416 - 2.800$$

$$= 0.3416$$

$$\sin 2.800 = \sin 0.3416$$

$$= 0.3350$$

Example 5-5.4 Find $\cos 10.210$.

Solution $\qquad t = 10.210$

$$t = t_1 + 6.283$$

$$t_1 = 10.210 - 6.283$$

$$t_1 = 3.927$$

t_1 is represented by a point in quadrant III, hence

$$t_r = t_1 - 3.1416$$

$$t_r = 3.927 - 3.1416$$

$$= 0.785$$

$$\cos 10.210 = -\cos 0.785$$

$$= -0.7072$$

Example 5-5.5 Find csc 24.843.

$$t = 24.843$$

$$t_1 = t - 2k\pi$$

$$= t - 3(6.283)$$

$$= t - 18.849$$

$$= 24.843 - 18.849$$

$$= 5.994$$

$$t_r = 6.283 - 5.994$$

$$= 0.289$$

t_1 is represented by a point in quadrant IV, hence

$$\csc 24.843 = -\csc 0.289$$

Hence csc 24.843 is approximately -3.500.

Exercise 5-5.1   Use Table C-2 to evaluate the following:

1   sec 0.91   **2**   cot 1   **3**   sin 2.45

4   cos 4.053   **5**   tan 5.807   **6**   csc 10.31

7   sec 15.42   **8**   cot 22.43   **9**   sin 122.78

10   cos 214.33

Review Exercise   **1**   Use the expansion formulas to evaluate:

(*a*)   sin 4   (*b*)   cos 2

2   Give the sine, cosine, and tangent of t if t is represented on the unit circle by the point:

(*a*) $P\left(\dfrac{1}{2}, \dfrac{\sqrt{3}}{2}\right)$ (*b*) $P(r,s)$

(*c*) $P\left(\dfrac{1}{\sqrt{2}}, -\dfrac{1}{\sqrt{2}}\right)$ (*d*) $P(0.9048, 0.4259)$

3   Give from memory the sine and cosine of the following:

(*a*) π (*b*) $\dfrac{3\pi}{2}$ (*c*) $\dfrac{7\pi}{6}$ (*d*) $\dfrac{5\pi}{3}$

4   Find a second number which has the same functional value as:

(*a*) $\sin \dfrac{4\pi}{3}$ (*b*) $\cos \dfrac{5\pi}{6}$ (*c*) $\tan \dfrac{7\pi}{6}$ (*d*) $\sec \dfrac{2\pi}{3}$

5   Use Table C-2 to find:

(*a*)   sec 0.25   (*b*)   csc 3.682   (*c*)   tan 28.75   (*d*)   sin 13.21

Answers

Exercise 5-2.1

1 $\sin 0.30 = 0.30 - \dfrac{0.30^3}{3!} + \dfrac{0.30^5}{5!} - \dfrac{0.30^7}{7!}$

$= 0.30 - \dfrac{0.0270}{6} + \dfrac{0.0024}{120} - \dfrac{0.000729}{5040}$

$= 0.30 - 0.0045 + 0.00002$

$= 0.30002 - 0.0045$

$= 0.2955$

2 $\cos 0.80 = 1 - \dfrac{0.80^2}{2!} + \dfrac{0.80^4}{4!} - \dfrac{0.80^6}{6!}$

$= 1 - 0.32 + \dfrac{0.4096}{24} - \dfrac{0.26214}{720}$

$= 1 - 0.32 + 0.017066 - 0.00036$

$= 1.017066 - 0.3204$

$= 0.6967$

3 $\sin 1.20 = 1.20 - \dfrac{(1.2)^3}{3!} + \dfrac{(1.2)^5}{5!}$

$= 1.20 - 0.2880 + 0.020736$

$= 1.2207 - 0.2880$

$= 0.9327$

$\cos 1.20 = 1. - \dfrac{(1.2)^2}{2} + \dfrac{(1.2)^4}{24} - \dfrac{(1.2)^6}{720}$

$= 1 - 0.720 + 0.0864 - 0.0041$

$= 1.0864 - 0.724$

$= 0.3624$

$\tan 1.20 = \dfrac{0.9327}{0.3624}$

$= 2.5737$

Exercise 5-3.1

1 $\sin t = \dfrac{1}{\sqrt{2}}$ $\qquad \tan t = 1 \qquad \sec t = \sqrt{2}$

$\cos t = \dfrac{1}{\sqrt{2}}$ $\qquad \cot t = 1 \qquad \csc t = \sqrt{2}$

2 $\sin t = 0 \qquad\qquad \tan t = 0 \qquad\quad \sec t = 1$

$\cos t = 1 \qquad\qquad \cot t = \text{undef.} \quad \csc t = \text{undef.}$

3 $\sin t = \dfrac{1}{\sqrt{2}}$ $\qquad \tan t = -1 \qquad \sec t = -\sqrt{2}$

$\cos t = -\dfrac{1}{\sqrt{2}}$ $\quad \cot t = -1 \qquad \csc t = \sqrt{2}$

4 $\sin t = 0$ $\tan t = 0$ $\sec t = -1$
    $\cos t = -1$ $\cot t = \text{undef.}$ $\csc t = \text{undef.}$

5 $\sin t = \dfrac{1}{2}$ $\tan t = \dfrac{1}{\sqrt{3}}$ $\sec t = \dfrac{2}{\sqrt{3}}$

    $\cos t = \dfrac{\sqrt{3}}{2}$ $\cot t = \sqrt{3}$ $\csc t = 2$

6 $\sin t = -\dfrac{\sqrt{3}}{2}$ $\tan t = \sqrt{3}$ $\sec t = -2$

    $\cos t = -\dfrac{1}{2}$ $\cot t = \dfrac{1}{\sqrt{3}}$ $\csc t = -\dfrac{2}{\sqrt{3}}$

Exercise 5-4.1 **1** $\frac{1}{2}$ **2** Undefined   **3** -1 **4** $\dfrac{2}{\sqrt{3}}$

5 $-\dfrac{2}{\sqrt{3}}$ **6** $-\frac{1}{2}$ **7** 0 **8** $\dfrac{\pi}{3}$

9 $\dfrac{5\pi}{3}$ **10** $\dfrac{5\pi}{3}$ **11** $\dfrac{\pi}{4}$ **12** $\dfrac{\pi}{4}$

Exercise 5-5.1 **1** 1.629 **2** 0.6421 **3** 0.6382 **4** -0.6129
    5 -0.5156 **6** -1.272 **7** -1.041 **8** 2.124
    9 -0.2571 **10** 0.7637

Review Exercise **1** (*a*) $\sin 4 = 4 - \dfrac{4^3}{3!} + \dfrac{4^5}{5!} - \dfrac{4^7}{7!} + \dfrac{4^9}{9!}$

$= 4 - \dfrac{64}{6} + \dfrac{1024}{120} - \dfrac{16{,}384}{5040} + \dfrac{262{,}144}{362{,}880}$

$= 4 - 10.667 + 8.5333 - 3.2508 + 0.7224$

$= 13.2557 - 13.9178$

$= -0.6621$

(*b*) $\cos 2 = 1 - \dfrac{2^2}{2!} + \dfrac{2^4}{4!} - \dfrac{2^6}{6!} + \dfrac{2^8}{8!}$

$= 1 - 2 + 0.66667 - 0.08889 + 0.00635$

$= -0.4158$

2 (*a*) $\sin t = \dfrac{\sqrt{3}}{2}$ (*b*) $\sin t = s$ (*c*) $\sin t = -\dfrac{1}{\sqrt{2}}$ (*d*) $\sin t = 0.4259$

$\cos t = \dfrac{1}{2}$ $\cos t = r$ $\cos t = \dfrac{1}{\sqrt{2}}$ $\cos t = 0.9048$

$\tan t = \sqrt{3}$ $\tan t = \dfrac{s}{r}$ $\tan t = -1$ $\tan t = 0.4707$

3　(a)　$\sin \pi = 0$

　　　　$\cos \pi = -1$

　　(c)　$\sin \dfrac{7\pi}{6} = -\dfrac{1}{2}$

　　　　$\cos \dfrac{7\pi}{6} = -\dfrac{\sqrt{3}}{2}$

　　(b)　$\sin \dfrac{3\pi}{2} = -1$

　　　　$\cos \dfrac{3\pi}{2} = 0$

　　(d)　$\sin \dfrac{5\pi}{3} = -\dfrac{\sqrt{3}}{2}$

　　　　$\cos \dfrac{5\pi}{3} = \dfrac{1}{2}$

4　(a)　$\sin \dfrac{4\pi}{3} = \sin \dfrac{5\pi}{3}$

　　(c)　$\tan \dfrac{7\pi}{6} = \tan \dfrac{\pi}{6}$

　　(b)　$\cos \dfrac{5\pi}{6} = \cos \dfrac{7\pi}{6}$

　　(d)　$\sec \dfrac{2}{3}\pi = \sec \dfrac{4}{3}\pi$

5　(a)　$\sec 0.25 = 1.032$
　　(b)　$\csc 3.682 = -1.945$
　　(c)　$\tan 28.75 = 0.5206$
　　(d)　$\sin 13.21 = 0.5972$

Graphing Trigonometric Functions

6

6/Graphing Trigonometric Functions

6-1 Introduction   In your study of algebra you learned how to graph various algebraic functions including linear functions, quadratic functions, and absolute value functions. The six trigonometric functions are part of a class of functions known as "transcendental functions," and in this chapter you will learn how to sketch the graphs of these functions.

Before studying graphs of the trigonometric functions it is beneficial to recall some facts about the domain and range of these functions. Table 6-1.1 summarizes this information. You may wish to review Appendix A-2.

Table 6-1.1

Function	Domain	Range
sine: $y = \sin x$*	$\{x \mid x \text{ is an angle}\}$	$\{y \mid -1 \leq y \leq 1\}$
cosine: $y = \cos x$	$\{x \mid x \text{ is an angle}\}$	$\{y \mid -1 \leq y \leq 1\}$
tangent: $y = \tan x$	$\{x \mid x \in R, x \neq \dfrac{(2n + 1)\pi}{2}, n \in Z\}$	$\{y \mid y \in R\}$
cotangent: $y = \cot x$	$\{x \mid x \in R, x \neq n\pi, n \in Z\}$	$\{y \mid y \in R\}$
secant: $y = \sec x$	$\{x \mid x \in R, x \neq \dfrac{(2n + 1)\pi}{2}, n \in Z\}$	$\{y \mid y \geq 1, y \leq -1\}$
cosecant: $y = \csc x$	$\{x \mid x \in R, x \neq n\pi, n \in Z\}$	$\{y \mid y \geq 1, y \leq -1\}$

*Note that x or θ or any other variable may be used to represent the measure of an angle.

We will now consider the graphs of each function separately, pointing out useful reference points and giving specific steps to follow in order to obtain sketches of the curves.

6-2 Graphing Sine Functions   Consider the function $y = \sin x$. You could obtain the sketch shown in Figure 6-2.1 by plotting many points, where $0° \leq x \leq 360°$ (see Table 6-2.1).

Table 6-2.1

x	0°	30°	45°	60°	90°	120°	135°	150°	180°
$\sin x$	0°	0.5	0.7	0.87	1	0.87	0.7	0.5	0

x	210°	225°	240°	270°	300°	315°	330°	360°
$\sin x$	−0.5	−0.7	−0.87	−1	−0.87	−0.7	−0.5	0

Figure 6-2.1

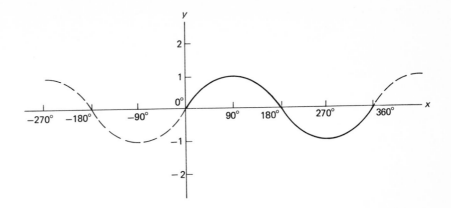

Notice that for the graph of the function $y = \sin x$ the curve repeats itself every 360°. That is, one complete cycle of the curve occurs as x changes from 0° to 360°. We say the *period* of $y = \sin x$ is 360°, or 2π. The curve can be extended in both directions, and its shape will look just like that already sketched. The dotted line in Figure 6-2.1 denotes an extension of the curve. Note that the maximum value is 1 and the minimum value is -1. One half of the distance from the minimum value to the maximum value is called the *amplitude*. The graph crosses the x axis when x takes the values 0, π, and 2π.

The graphs we must sketch often involve more complicated equations. Instead of $y = \sin x$ we sometimes encounter equations of the form $y = a \sin (px + s) + h$, where a, p, s, and h are real numbers. Before attempting to graph such an equation we will see how each of the numbers a, p, s, and h affects the graph.

It will be much easier to sketch the graphs if you remember how each constant alters the basic curve.

The constant a will affect the amplitude of the function. Hence, the amplitude of the graph of $y = a \sin (px + s) + h$ will differ from the amplitude of $y = \sin x$ by a factor of $|a|$. In essence, the amplitude will increase or decrease according to $|a|$. Example 6-2.1 illustrates how a affects the amplitude. Note the comparison with $y = \sin x$ which is represented in Figure 6-2.2 as a dotted line.

Example 6-2.1 Sketch the graph of $y = 2 \sin x$.

Solution Make a table of reference points using the values of x that cause the y values to become maximum, minimum, and zero. (See Table 6-2.2.) Use these coordinates to sketch one cycle of the graph.

Table 6-2.2

x	0	$\frac{\pi}{2}$	π	$\frac{3\pi}{2}$	2π
$\sin x$	0	1	0	-1	0
$y = 2 \sin x$	0	2	0	-2	0

Figure 6-2.2

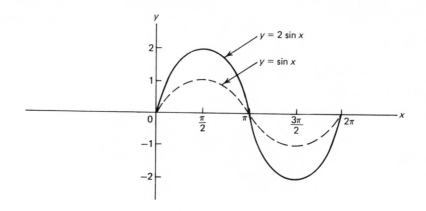

The constant p will affect the period of the curve $y = a \sin (px + s) + h$ in such a way that its period will differ from that of $y = \sin x$ by a factor of $\frac{1}{|p|}$. In Example 6-2.2 we will compare the graphs of $y = \sin 2x$ and $y = \sin x$ to show how the period of the function changes when $p = 2$.

Example 6-2.2 Sketch the graph of $y = \sin 2x$.

Solution Make a table of reference points using the values of x that cause the y coordinates to become maximum, minimum, and zero. (See Table 6-2.3.) Use these coordinates to sketch one cycle of the graph.

Table 6-2.3

x	0	$\frac{\pi}{4}$	$\frac{\pi}{2}$	$\frac{3\pi}{4}$	π	
$2x$	0	$\frac{\pi}{2}$	π	$\frac{3\pi}{2}$	2π	← Fill in first
$y = \sin 2x$	0	1	0	-1	0	

Note that by first filling in the middle line of Table 6-2.3 we can be assured of choosing values of x that give maximum, minimum, and zero values of y.

Figure 6-2.3

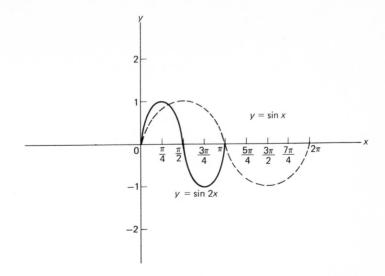

The constants s and p together determine the shift of a curve along the x axis. The curve of $y = a \sin (px + s)$ will shift to the left by a factor of $\left|\dfrac{s}{p}\right|$ if $\dfrac{s}{p} > 0$, and to the right by a factor of $\left|\dfrac{s}{p}\right|$ if $\dfrac{s}{p} < 0$. In Example 6-2.3 we will see how this works. Figure 6-2.4 shows a comparison of a shifted graph with the graph of $y = \sin x$.

Example 6-2.3 Sketch one cycle of the graph of $y = \sin\left(x + \dfrac{\pi}{2}\right)$.

Solution Here we see that $p = 1$ and $s = \dfrac{\pi}{2}$; hence, $\left|\dfrac{s}{p}\right| = \left|\dfrac{\pi/2}{1}\right| = \left|\dfrac{\pi}{2}\right|$. Since $\dfrac{\pi}{2} > 0$, our curve will shift to the left $\dfrac{\pi}{2}$ units. Make a table of reference points (Table 6-2.4).

Table 6-2.4

x	$-\dfrac{\pi}{2}$	0	$\dfrac{\pi}{2}$	π	$\dfrac{3\pi}{2}$	
$x + \dfrac{\pi}{2}$	0	$\dfrac{\pi}{2}$	π	$\dfrac{3\pi}{2}$	2π	← Fill in first
$y = \sin\left(x + \dfrac{\pi}{2}\right)$	0	1	0	-1	0	

Use these points to sketch the curve.

Figure 6-2.4

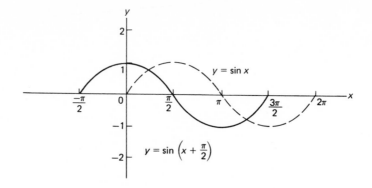

The final change in the graph of $y = a \sin (px + s) + h$ occurs when h is not zero. If $h > 0$, the graph is raised h units above the x axis, and if $h < 0$, the graph is lowered h units below the x axis. Such a shift is shown in Example 6-2.4 and Figure 6-2.5.

Example 6-2.4 Sketch one cycle of the graph $y = \sin x + 1$.

Solution Make a table of reference points (Table 6-2.5).

Table 6-2.5

x	0	$\dfrac{\pi}{2}$	π	$\dfrac{3\pi}{2}$	2π	← Fill in first
$\sin x$	0	1	0	-1	0	
$y = \sin x + 1$	1	2	1	0	1	

Use these points to sketch the curve.

Figure 6-2.5

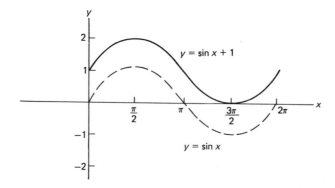

In Examples 6-2.1, 6-2.2, 6-2.3, and 6-2.4 we used reference points to help us sketch the desired graphs. The "reference" points are those points that tell us where the graph reaches a maximum, a minimum, or a zero value. Table 6-2.6 summarizes the reference points that should be used for graphing the sine function.

Table 6-2.6

Reference points for
$y = \sin x$

x	0	$\dfrac{\pi}{2}$	π	$\dfrac{3\pi}{2}$	2π
y	0	1	0	-1	0

Now we are ready to learn a procedure for sketching the graph of $y = a \sin (px + s) + h$. Here are the steps:

STEP 1 If $s \neq 0$, factor the p out of $(px + s)$, giving $y = a \sin p\left(x + \dfrac{s}{p}\right) + h$.

STEP 2 Multiply 2π by $\dfrac{1}{|p|}$ to get the period. The period of $y = a \sin (px + s) + h$ is $\dfrac{2\pi}{|p|}$.

STEP 3 Construct a reference table for $y = a \sin p\left(x + \dfrac{s}{p}\right) + h$ as illustrated in Table 6-2.7. Do this in the following steps:
  (*a*) Make six columns.
  (*b*) In the left-hand column begin labeling the rows with x, and build one step at a time until the last row is labeled y.
  (*c*) Use the reference points given in Table 6-2.6 to fill in rows $p\left(x + \dfrac{s}{p}\right)$ and $\sin p\left(x + \dfrac{s}{p}\right)$. Always fill in these rows first and second respectively.

Table 6-2.7

Reference table for
$y = a \sin (px + s) + h$

x	$0 - \dfrac{s}{p}$	$\dfrac{\pi/2}{p} - \dfrac{s}{p}$	$\dfrac{\pi}{p} - \dfrac{s}{p}$	$\dfrac{3\pi/2}{p} - \dfrac{s}{p}$	$\dfrac{2\pi}{p} - \dfrac{s}{p}$
$x + \dfrac{s}{p}$	0	$\dfrac{\pi/2}{p}$	$\dfrac{\pi}{p}$	$\dfrac{3\pi/2}{p}$	$\dfrac{2\pi}{p}$
$p\left(x + \dfrac{s}{p}\right)$	0	$\dfrac{\pi}{2}$	π	$\dfrac{3\pi}{2}$	2π
$\sin p\left(x + \dfrac{s}{p}\right)$	0	1	0	-1	0
$a \sin p\left(x + \dfrac{s}{p}\right)$	$a(0)$	$a(1)$	$a(0)$	$a(-1)$	$a(0)$
$y = a \sin p(x + s) + h$	$0 + h$	$a + h$	$0 + h$	$-a + h$	$0 + h$

← Fill in this row first from Table 6-2.6

(d)   Now work backward to the *x* values and forward to the *y* values.

STEP 4   Sketch the graph using your knowledge of the basic shape of $y = \sin x$ and the ordered pairs (x,y) as given in the first and last entries in each column.

The following examples will illustrate how the technique works on different equations of the form $y = a \sin (px + s) + h$.

Example 6-2.5   Sketch the graph of $y = 3 \sin 2x$.

Solution   **STEP 1**   No need to factor out p since $s = 0$. Note that $p = 2$.

STEP 2   Find the period of $y = 3 \sin 2x$ by multiplying $\dfrac{1}{|2|} \cdot 2\pi$. Thus the period is π. The factor 2 changes the period from 2π to π.

STEP 3   Construct a reference table as shown in Table 6-2.8.

Table 6-2.8
Reference table for
$y = 3 \sin 2x$

x	0	$\dfrac{\pi}{4}$	$\dfrac{\pi}{2}$	$\dfrac{3\pi}{4}$	π	
$2x$	0	$\dfrac{\pi}{2}$	π	$\dfrac{3\pi}{2}$	2π	← Fill in this row first.
$\sin 2x$	0	1	0	-1	0	
$y = 3 \sin 2x$	0	3	0	-3	0	

STEP 4   Sketch the graph using the ordered pairs formed by the first and the last entries in each column: $(0,0)$, $\left(\dfrac{\pi}{4}, 3\right)$, $\left(\dfrac{\pi}{2}, 0\right)$, $\left(\dfrac{3\pi}{4}, -3\right)$, $(\pi, 0)$. See Figure 6-2.6.

Figure 6-2.6

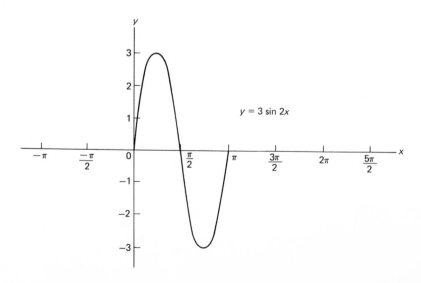

$y = 3 \sin 2x$

Example 6-2.6 Sketch the graph of $y = \frac{3}{2} \sin (3x - \pi)$.

Solution **STEP 1** Factor out the 3 giving $y = \frac{3}{2} \sin 3\left(x - \frac{\pi}{3}\right)$.

STEP 2 Find the period by multiplying $\frac{1}{|3|} \cdot 2\pi = \frac{2\pi}{3}$.

STEP 3 Construct a reference table as shown in Table 6-2.9.

Table 6-2.9
Reference table for
$y = \frac{3}{2} \sin (3x - \pi)$

	$\frac{\pi}{3}$	$\frac{\pi}{2}$	$\frac{2\pi}{3}$	$\frac{5\pi}{6}$	π	
x						
$x - \dfrac{\pi}{3}$	0	$\dfrac{\pi}{6}$	$\dfrac{\pi}{3}$	$\dfrac{\pi}{2}$	$\dfrac{2\pi}{3}$	
$3\left(x - \dfrac{\pi}{3}\right)$	0	$\dfrac{\pi}{2}$	π	$\dfrac{3\pi}{2}$	2π	← Fill in this row first
$\sin 3\left(x - \dfrac{\pi}{3}\right)$	0	1	0	-1	0	
$y = \dfrac{3}{2} \sin 3\left(x - \dfrac{\pi}{3}\right)$	0	$\dfrac{3}{2}$	0	$-\dfrac{3}{2}$	0	

STEP 4 Use the ordered pairs formed by the first and the last entries in each column to sketch one cycle of the graph: $\left(\frac{\pi}{3}, 0\right), \left(\frac{\pi}{2}, \frac{3}{2}\right), \left(\frac{2\pi}{3}, 0\right), \left(\frac{5\pi}{6}, -\frac{3}{2}\right),$ $(\pi, 0)$. See Figure 6-2.7.

Figure 6-2.7

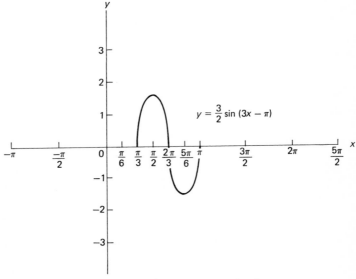

$y = \dfrac{3}{2} \sin (3x - \pi)$

Example 6-2.7 Sketch one cycle of the graph $y = \frac{3}{4} \sin \left(\frac{x + \pi}{2}\right) + 1$.

Solution **STEP 1** Factor out the $\frac{1}{2}$, giving

$$y = \tfrac{3}{4} \sin \tfrac{1}{2} (x + \pi) + 1$$

STEP 2 Find the period.

STEP 3 Construct a reference table as in Table 6-2.10.

Table 6-2.10

Reference table for
$y = \frac{3}{4} \sin \left(\frac{x + \pi}{2}\right) + 1$

x	$-\pi$	0	π	2π	3π	
$x + \pi$	0	π	2π	3π	4π	
$\frac{1}{2}(x + \pi)$	0	$\frac{\pi}{2}$	π	$\frac{3\pi}{2}$	2π	← Fill in this row first
$\sin \frac{1}{2}(x + \pi)$	0	1	0	-1	0	
$\frac{3}{4} \sin \frac{1}{2}(x + \pi)$	0	$\frac{3}{4}$	0	$-\frac{3}{4}$	0	
$y = \frac{3}{4} \sin \frac{1}{2}(x + \pi) + 1$	1	$\frac{7}{4}$	1	$\frac{1}{4}$	1	

STEP 4 Sketch one cycle of the graph using the ordered pairs formed by the first and last entries in each column: $(-\pi, 1)$, $(0, \frac{7}{4})$, $(\pi, 1)$, $(2\pi, \frac{1}{4})$, $(3\pi, 1)$. See Figure 6-2.8.

Figure 6-2.8

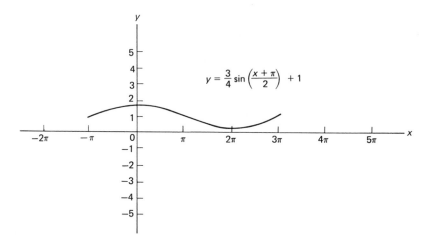

$$y = \frac{3}{4} \sin \left(\frac{x + \pi}{2}\right) + 1$$

Exercise 6-2.1 In problems 1–5 of the form $y = a \sin (px + s) + h$, give the values of a, p, s, and h and find the amplitude, period, and phase shift along the x axis and vertical shift relative to the x axis.

1 $y = 2 \sin x$

2 $y = -\frac{1}{3} \sin (2x + \pi)$

3 $y = \frac{1}{5} \sin (\frac{1}{2}x + 2\pi) + 4$　　　　**4** $y = 5 \sin (-2x + \pi) - 2$

5 $y = -2 \sin (\frac{1}{3}x + 3\pi) + \pi$

In problems 6–15 use the techniques of this section to sketch the graph of one cycle of the given function.

6 $y = 3 \sin x$　　　　　　　　　　**7** $y = 2 \sin \frac{1}{2}x$

8 $y = \frac{1}{2} \sin 2x$　　　　　　　　**9** $y = \frac{1}{8} \sin (3x + \pi)$

10 $y = -3 \sin 4x$　　　　　　　**11** $y = \sin (3x + \pi)$

12 $y = \sin \dfrac{2x}{3} + 2$　　　　　**13** $y = 2 \sin \left(\dfrac{x}{3} + \pi\right) + 1$

14 $y = -\sin (x + 2\pi) + 2$　　　**15** $y = \frac{1}{2} \sin (2x + 2\pi) - 1$

6-3 Graphing Cosine Functions

The graph of $y = \cos x$ is the same as that of $y = \sin \left(x + \dfrac{\pi}{2}\right)$. This tells us that the graph of $y = \cos x$ has the same shape, amplitude, and period as the graph of $y = \sin x$. In Figure 6-3.1 we have used the points shown in Table 6-3.1 to sketch the graph of $y = \cos x$ through one cycle from $x = 0°$ to $x = 360°$. The dotted curve shows the extension of the graph beyond one period.

Table 6-3.1

θ	0°	30°	45°	60°	90°	120°	135°	150°	180°
$\cos \theta$	1	0.87	0.7	0.5	0	−0.5	−0.7	−0.87	−1

θ	210°	225°	240°	270°	300°	315°	330°	360°
$\cos \theta$	−0.87	−0.7	−0.5	0	0.5	0.7	0.87	1

Figure 6-3.1

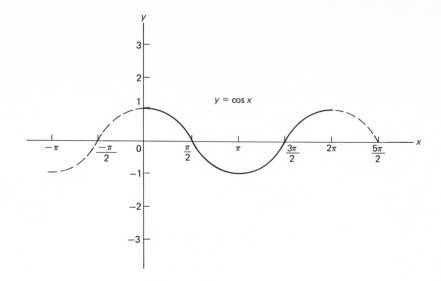

The method for graphing $y = a \sin (px + s) + h$ works equally well for graphing $y = a \cos (px + s) + h$. Remember that a affects the amplitude, p affects the period, $\dfrac{s}{p}$ affects the shift to the left or right, and h affects the position of the graph up or down with respect to the x axis.

Table 6-3.2 summarizes the reference points for one cycle of $y = \cos x$. We will use it to help sketch functions of the form $y = a \cos (px + s) + h$ in the same way we used the table of reference points for $y = \sin x$.

Table 6-3.2

Reference table for $y = \cos x$, $0 \leq x \leq 2\pi$

x	0	$\dfrac{\pi}{2}$	π	$\dfrac{3\pi}{2}$	2π
y	1	0	-1	0	1

The following examples will illustrate how this technique may be used to sketch the graph of equations of the form $y = a \cos (px + s) + h$.

Example 6-3.1   Sketch one cycle of the graph for $y = 2 \cos 3x$.

Solution   **STEP 1**   We need not factor out 3 since $s = 0$.

STEP 2   Find the period.

$$\text{Period} = \frac{1}{|3|} \cdot 2\pi = \frac{2\pi}{3}$$

STEP 3   Construct a reference table, as shown in Table 6-3.3.

Table 6-3.3

Reference table for
$y = 2 \cos 3x$

x	0	$\dfrac{\pi}{6}$	$\dfrac{\pi}{3}$	$\dfrac{\pi}{2}$	$\dfrac{2\pi}{3}$	
$3x$	0	$\dfrac{\pi}{2}$	π	$\dfrac{3\pi}{2}$	2π	← Fill in this row first
$\cos 3x$	1	0	−1	0	1	
$y = 2 \cos 3x$	2	0	−2	0	2	

STEP 4 Sketch the graph using the ordered pairs $(0,2)$, $\left(\dfrac{\pi}{6},0\right)$, $\left(\dfrac{\pi}{3},-2\right)$, $\left(\dfrac{\pi}{2},0\right)$, and $\left(\dfrac{2\pi}{3},2\right)$.

Figure 6-3.2

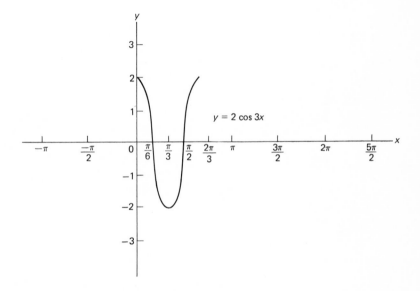

Example 6-3.2 Sketch the graph of one cycle of $y = \frac{1}{3} \cos \left(\frac{1}{2}x - 1\right)$.

Solution **STEP 1** Factor out $\frac{1}{2}$ to give

$$y = \tfrac{1}{3} \cos \tfrac{1}{2}(x - 2)$$

STEP 2 Find the period.

$$\text{Period} = \frac{1}{\left|\frac{1}{2}\right|} \cdot 2\pi = 4\pi$$

STEP 3 Construct a reference table, as in Table 6-3.4.

Table 6-3.4

Reference table for
$y = \frac{1}{3} \cos (\frac{1}{2}x - 1)$

x	2	$\pi + 2$	$2\pi + 2$	$3\pi + 2$	$4\pi + 2$	
$x - 2$	0	π	2π	3π	4π	
$\frac{1}{2}(x - 2)$	0	$\frac{\pi}{2}$	π	$\frac{3\pi}{2}$	2π	← Fill in this row first
$\cos \frac{1}{2}(x - 2)$	1	0	-1	0	1	
$y = \frac{1}{3} \cos \frac{1}{2}(x - 2)$	$\frac{1}{3}$	0	$-\frac{1}{3}$	0	$\frac{1}{3}$	

STEP 4 Sketch one cycle of $y = \frac{1}{3} \cos (\frac{1}{2}x - 1)$ by using the ordered pairs $(2,\frac{1}{3})$, $(\pi + 2,0)$, $(2\pi + 2,-\frac{1}{3})$, $(3\pi + 2,0)$, and $(4\pi + 2,\frac{1}{3})$.

Figure 6-3.3

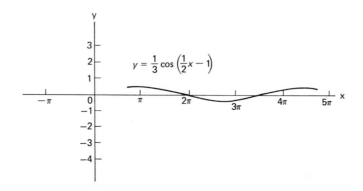

Example 6-3.3 Sketch one cycle of the graph of $y = 2 \cos (2x - 2\pi) - 2$.

Solution **STEP 1** Factor out 2, giving $y = 2 \cos 2(x - \pi) - 2$.

STEP 2 Find the period.

$$\text{Period} = \frac{1}{|2|} \cdot 2\pi = \pi$$

STEP 3 Construct a reference table for $y = 2 \cos (2x - 2\pi) - 2$.

Table 6-3.5

Reference table for
$y = 2 \cos 2(x - \pi) - 2$

x	π	$\frac{5\pi}{4}$	$\frac{3\pi}{2}$	$\frac{7\pi}{4}$	2π	
$x - \pi$	0	$\frac{\pi}{4}$	$\frac{\pi}{2}$	$\frac{3\pi}{4}$	π	
$2(x - \pi)$	0	$\frac{\pi}{2}$	π	$\frac{3\pi}{2}$	2π	← Fill in this row first
$\cos 2(x - \pi)$	1	0	-1	0	1	
$2 \cos 2(x - \pi)$	2	0	-2	0	2	
$y = 2 \cos 2(x - \pi) - 2$	0	-2	-4	-2	0	

STEP 4   Sketch one cycle of $y = 2 \cos 2(x - \pi) - 2$ using the ordered pairs $(\pi, 0)$, $\left(\dfrac{5\pi}{4}, -2\right)$, $\left(\dfrac{3\pi}{2}, -4\right)$, $\left(\dfrac{7\pi}{4}, -2\right)$, $(2\pi, 0)$.

Figure 6-3.4

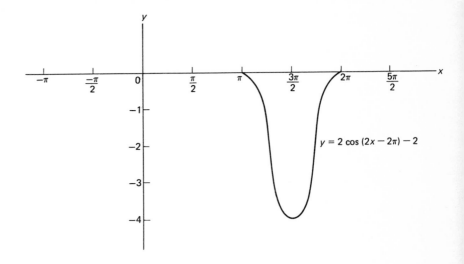

$y = 2 \cos (2x - 2\pi) - 2$

Exercise   6-3.1   In problems 1–5, which are of the form $y = a \cos (px + s) + h$, give the values of $a, p, s,$ and h and find the amplitude, period, phase shift along the x axis, and vertical shift relative to the x axis.

1 $y = \cos 3x$

2 $y = 2 \cos 4x$

3 $y = \tfrac{1}{2} \cos (x + \pi)$

4 $y = -2 \cos \left(2x + \dfrac{\pi}{2}\right)$

5 $y = 3 \cos (3x - \pi) + 2$

In problems 6–15, use the techniques of this section to sketch one cycle of the given function.

6 $y = 3 \cos x$ **7** $y = \cos (2x + \pi)$

8 $y = \frac{1}{2} \cos (2x - \pi)$ **9** $y = -2 \cos (\frac{1}{2}x + 2\pi)$

10 $y = \frac{3}{2} \cos \left(x + \dfrac{\pi}{2} \right)$ **11** $y = -3 \cos (3x + 1)$

12 $y = 2 \cos (x + 2\pi) + 1$ **13** $y = 3 \cos (4x - 2\pi) - 2$

14 $y = -\frac{3}{4} \cos (\frac{1}{2}x + \pi) + 1$ **15** $y = 4 \cos \left(-2x + \dfrac{\pi}{2} \right) - 1$

6-4 Graphing Tangent Functions

The method we have developed for sketching the graphs of $y = \sin x$ and $y = \cos x$ will work equally well when applied to $y = \tan x$. We will, however, need to make several minor adjustments in our technique. First let us look at the graph of $y = \tan x$, as shown in Figure 6-4.1.

Table 6-4.1

x	$-90°$	$-60°$	$-45°$	$-30°$	$0°$	$30°$	$45°$	$60°$	$90°$
$\tan x$	Undef.	-1.7	-1	-0.6	0	0.6	1	1.7	Undef.

x	$120°$	$135°$	$150°$	$180°$	$210°$	$225°$	$240°$	$270°$
$\tan x$	-1.7	-1	-0.6	0	0.6	1	1.7	Undef.

Plotting the points from Table 6-4.1 and connecting them produce the graph for $y = \tan x$ shown in Figure 6-4.1.

Figure 6-4.1

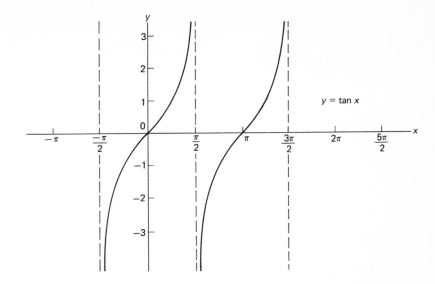

We see that one cycle of the graph of $y = \tan x$ occurs between $-\dfrac{\pi}{2}$ and $\dfrac{\pi}{2}$ and that another cycle occurs between $\dfrac{\pi}{2}$ and $\dfrac{3\pi}{2}$. Hence, the period of the function $y = \tan x$ is π, or $180°$. The graph crosses the x axis at integral multiples of π. It crosses the y axis at $x = 0$. The graph increases and decreases without bounds when x values approach odd multiples of $\dfrac{\pi}{2}$. Table 6-4.2 summarizes the reference points that we can use in sketching graphs of functions of the form $y = a \tan (px + s) + h$. Note that in the absence of maximum and minimum values we use two auxiliary points where $\tan x = 1$ and $\tan x = -1$.

Table 6-4.2

Reference table for $y = \tan x$

x	$-\dfrac{\pi}{2}$	$-\dfrac{\pi}{4}$	0	$\dfrac{\pi}{4}$	$\dfrac{\pi}{2}$
y	Undef.*	-1	0	1	Undef.†

*The function decreases without bounds.
†The function increases without bounds.

We now use the method developed in Sections 6-1 and 6-2 for graphing functions of the form $y = a \tan (px + s) + h$. The following examples illustrate the technique.

Example 6-4.1   Sketch one cycle of the graph of $y = 2 \tan 2x$.

Solution   **STEP 1**   We need not factor out 2 since $s = 0$.

STEP 2 Find the period:

$$\frac{1}{|2|} \cdot \pi = \frac{\pi}{2}$$

STEP 3 Construct a reference table for $y = 2 \tan 2x$.

Table 6-4.3

Reference table for $y = 2 \tan 2x$

x		$-\frac{\pi}{4}$	$-\frac{\pi}{8}$	0	$\frac{\pi}{8}$	$\frac{\pi}{4}$	
$2x$		$-\frac{\pi}{2}$	$-\frac{\pi}{4}$	0	$\frac{\pi}{4}$	$\frac{\pi}{2}$	← Fill in this row first
$\tan 2x$		Undef.	-1	0	1	Undef.	
$y = 2 \tan 2x$		Undef.	-2	0	2	Undef.	

STEP 4 Sketch one cycle of $y = 2 \tan 2x$ using the ordered pairs $\left(-\frac{\pi}{4}, \text{undefined}\right)$, $\left(-\frac{\pi}{8}, -2\right)$, $(0,0)$, $\left(\frac{\pi}{8}, 2\right)$, $\left(\frac{\pi}{4}, \text{undefined}\right)$.

Figure 6-4.2

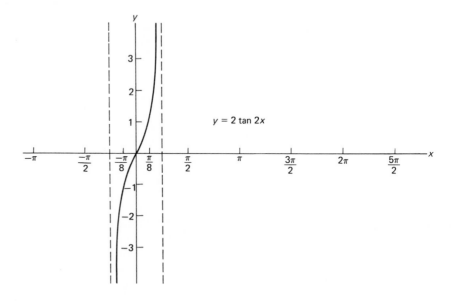

Example 6-4.2 Sketch one cycle of the graph of $y = \frac{1}{4} \tan\left(\frac{x - \pi}{2}\right)$.

Solution **STEP 1** Factor out $\frac{1}{2}$:

$$y = \frac{1}{4} \tan \frac{1}{2}(x - \pi).$$

STEP 2 Find the period.

$$\frac{1}{\left|\frac{1}{2}\right|} \cdot \pi = 2\pi$$

STEP 3 Construct a reference table.

Table 6-4.4
Reference table for
$y = \frac{1}{4}\tan\frac{1}{2}(x - \pi)$

x	0	$\frac{\pi}{2}$	π	$\frac{3\pi}{2}$	2π	
$x - \pi$	$-\pi$	$-\frac{\pi}{2}$	0	$\frac{\pi}{2}$	π	
$\frac{1}{2}(x - \pi)$	$-\frac{\pi}{2}$	$-\frac{\pi}{4}$	0	$\frac{\pi}{4}$	$\frac{\pi}{2}$	Fill in ← this row first
$\tan\frac{1}{2}(x - \pi)$	Undef.	-1	0	1	Undef.	
$y = \frac{1}{4}\tan\frac{1}{2}(x - \pi)$	Undef.	$-\frac{1}{4}$	0	$\frac{1}{4}$	Undef.	

STEP 4 Sketch one cycle of the graph of $y = \frac{1}{4}\tan\frac{1}{2}(x - \pi)$ using the ordered pairs $(0, \text{undefined})$, $\left(\frac{\pi}{2}, -\frac{1}{4}\right)$, $(\pi, 0)$, $\left(\frac{3\pi}{2}, \frac{1}{4}\right)$, $(2\pi, \text{undefined})$.

Figure 6-4.3

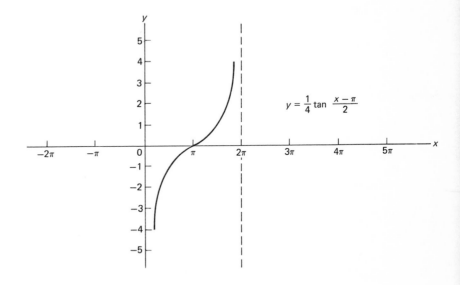

Example 6-4.3 Sketch one cycle of the graph of $y = 2\tan(2x - \pi) + 2$.

Solution **STEP 1** Factor out 2:

$$y = 2\tan 2\left(x - \frac{\pi}{2}\right) + 2$$

STEP 2 Find the period.

$$\frac{1}{|2|} \cdot \pi = \frac{\pi}{2}$$

STEP 3 Construct a reference table.

Table 6-4.5

Reference table for $y = 2 \tan 2\left(x - \frac{\pi}{2}\right) + 2$

x	$\frac{\pi}{4}$	$\frac{3\pi}{8}$	$\frac{\pi}{2}$	$\frac{5\pi}{8}$	$\frac{3\pi}{4}$	
$x - \frac{\pi}{2}$	$-\frac{\pi}{4}$	$-\frac{\pi}{8}$	0	$\frac{\pi}{8}$	$\frac{\pi}{4}$	
$2\left(x - \frac{\pi}{2}\right)$	$-\frac{\pi}{2}$	$-\frac{\pi}{4}$	0	$\frac{\pi}{4}$	$\frac{\pi}{2}$	← Fill in this row first
$\tan 2\left(x - \frac{\pi}{2}\right)$	Undef.	-1	0	1	Undef.	
$2 \tan 2\left(x - \frac{\pi}{2}\right)$	Undef.	-2	0	2	Undef.	
$y = 2 \tan 2\left(x - \frac{\pi}{2}\right) + 2$	Undef.	0	2	4	Undef.	

STEP 4 Sketch one cycle of $y = 2 \tan 2\left(x - \frac{\pi}{2}\right) + 2$ using the ordered pairs $\left(\frac{\pi}{4}, \text{undefined}\right)$, $\left(\frac{3\pi}{8}, 0\right)$, $\left(\frac{\pi}{2}, 2\right)$, $\left(\frac{5\pi}{8}, 4\right)$, $\left(\frac{3\pi}{4}, \text{undefined}\right)$.

Figure 6-4.4

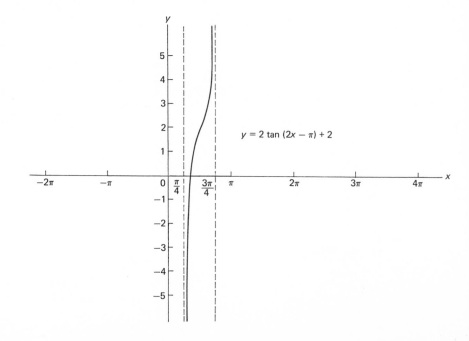

$y = 2 \tan (2x - \pi) + 2$

Exercise 6-4.1 In problems 1–5, which are of the form $y = a \tan (px + s) + h$, give the values of a, p, s, and h and find the period, phase shift, and vertical shift relative to the x axis.

1 $y = 3 \tan (2x + 3) + 5$

2 $y = \tan (x - 4) - 1$

3 $y = \frac{1}{2} \tan (2x - \pi)$

4 $y = 5 \tan \left(\frac{1}{3}x - \frac{\pi}{2} \right) + 1$

5 $y = 6 \tan (\frac{1}{5}x - \pi) - 2$

In problems 6–15, use the methods of this section to sketch one cycle of the graph of the following.

6 $y = 2 \tan 3x$ **7** $y = \frac{1}{2} \tan \frac{1}{4}x$

8 $y = -\frac{1}{4} \tan 6x$ **9** $y = 3 \tan (3x - \pi)$

10 $y = \frac{1}{2} \tan (2x - 2)$ **11** $y = 2 \tan (4x + 2\pi)$

12 $y = \tan \left(2x - \frac{\pi}{2} \right) + 1$ **13** $y = 2 \tan (x - \pi) - 3$

14 $y = -2 \tan \left(\frac{x}{4} + \pi \right) + 1$ **15** $y = 3 \tan \left(2x + \frac{\pi}{4} \right) - 2$

6-5 Graphing Cotangent Functions Following our previous pattern of presenting a table of values for each trigonometric function, we now do the same for $y = \cot x$. The graph of $y = \cot x$ is shown in Figure 6-5.1.

Table 6-5.1

x	0°		30°	45°	60°	90°	120°	135°	150°	180°
cot x	Undef.		1.7	1	0.6	0	−0.6	−1	−1.7	Undef.

x	210°	225°	240°	270°	300°	315°	330°	360°
cot x	1.7	1.0	0.6	0	−0.6	−1	−1.7	Undef.

Figure 6-5.1

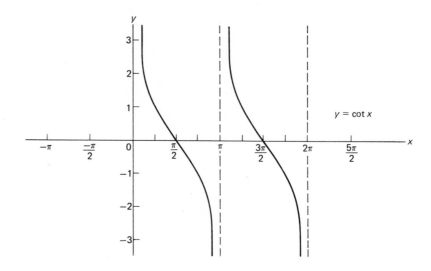

Figure 6-5.1 shows the period of $y = \cot x$ to be π. The graph crosses the x axis at odd-integer multiples of $\dfrac{\pi}{2}$ and becomes undefined at integral multiples of π.

We may apply our method of sketching graphs to equations of the form $y = a \cot (px + s) + h$ by using the table of reference points given in Table 6-5.2.

Table 6-5.2
Reference points for
$y = \cot x$

x	0	$\dfrac{\pi}{4}$	$\dfrac{\pi}{2}$	$\dfrac{3\pi}{4}$	π
y	Undef.	1	0	−1	Undef.

The following examples will illustrate the application of our method of sketching to the graphs of equations of the form $y = a \cot (px + s) + h$.

Example 6-5.1   Sketch one cycle of the graph of $y = 2 \cot 2x$.

Solution   **STEP 1**   Factor out p.   Unnecessary.

STEP 2   Find the period.

$$\frac{1}{|2|} \cdot \pi = \frac{\pi}{2}$$

STEP 3   Construct a reference table for $y = 2 \cot 2x$.

Table 6-5.3

Reference table for $y = 2 \cot 2x$

x	0	$\frac{\pi}{8}$	$\frac{\pi}{4}$	$\frac{3\pi}{8}$	$\frac{\pi}{2}$	
$2x$	0	$\frac{\pi}{4}$	$\frac{\pi}{2}$	$\frac{3\pi}{4}$	π	← Fill in first
$\cot 2x$	Undef.	1	0	-1	Undef.	
$y = 2 \cot 2x$	Undef.	2	0	-2	Undef.	

STEP 4   Use the ordered pairs $(0, \text{undefined})$, $\left(\frac{\pi}{8}, 2\right)$, $\left(\frac{\pi}{4}, 0\right)$, $\left(\frac{3\pi}{8}, -2\right)$, and $\left(\frac{\pi}{2}, \text{undefined}\right)$ to sketch one cycle of $y = 2 \cot 2x$.

Figure 6-5.2

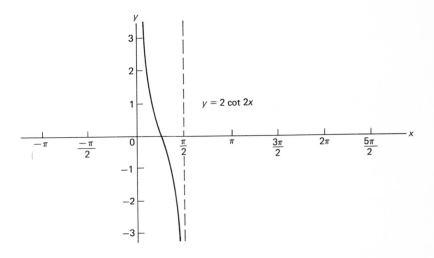

Example 6-5.2   Sketch one cycle of the graph of $y = \frac{1}{2} \cot \frac{1}{3}x$.

Solution   **STEP 1**   Factor out p.   Unnecessary.

STEP 2   Find the period.

$$\frac{1}{\left|\frac{1}{3}\right|} \cdot \pi = 3\pi$$

STEP 3   Construct a reference table for $y = \frac{1}{2} \cot \frac{1}{3}x$.

Table 6-5.4

Reference table for
$y = \frac{1}{2}\cot\frac{1}{3}x$

x	0	$\frac{3\pi}{4}$	$\frac{3\pi}{2}$	$\frac{9\pi}{4}$	3π	
$\frac{1}{3}x$	0	$\frac{\pi}{4}$	$\frac{\pi}{2}$	$\frac{3\pi}{4}$	π	← Fill in first
$\cot\frac{1}{3}x$	Undef.	1	0	-1	Undef.	
$y = \frac{1}{2}\cot\frac{1}{3}x$	Undef.	$\frac{1}{2}$	0	$-\frac{1}{2}$	Undef.	

STEP 4   Sketch one cycle of $y = \frac{1}{2}\cot\frac{1}{3}x$ using the ordered pairs $(0,\text{unde-fined})$, $\left(\frac{3\pi}{4},\frac{1}{2}\right)$, $\left(\frac{3\pi}{2},0\right)$, $\left(\frac{9\pi}{4},-\frac{1}{2}\right)$, and $(3\pi,\text{undefined})$.

Figure 6-5.3

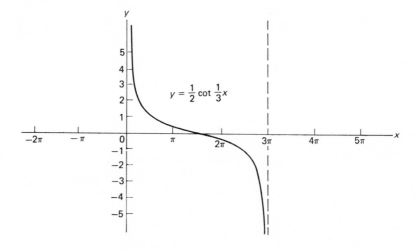

Example 6-5.3   Sketch one cycle of the graph of $y = 2\cot(2x - \pi) + 2$.

Solution   STEP 1   Factor out 2.

$$y = 2\cot 2\left(x - \frac{\pi}{2}\right) + 2$$

STEP 2 Find the period.

$$\frac{1}{|2|} \cdot \pi = \frac{\pi}{2}$$

STEP 3 Construct a reference table for $y = 2 \cot 2\left(x - \frac{\pi}{2}\right) + 2$.

Table 6-5.5

Reference table for
$y = 2 \cot 2\left(x - \frac{\pi}{2}\right) + 2$

x		$\frac{\pi}{2}$	$\frac{5\pi}{8}$	$\frac{3\pi}{4}$	$\frac{7\pi}{8}$	π	
$x - \frac{\pi}{2}$		0	$\frac{\pi}{8}$	$\frac{\pi}{4}$	$\frac{3\pi}{8}$	$\frac{\pi}{2}$	
$2\left(x - \frac{\pi}{2}\right)$		0	$\frac{\pi}{4}$	$\frac{\pi}{2}$	$\frac{3\pi}{4}$	π	← Fill in first
$\cot 2\left(x - \frac{\pi}{2}\right)$		Undef.	1	0	-1	Undef.	
$2 \cot 2\left(x - \frac{\pi}{2}\right)$		Undef.	2	0	-2	Undef.	
$y = 2 \cot 2\left(x - \frac{\pi}{2}\right) + 2$		Undef.	4	2	0	Undef.	

STEP 4 Use the ordered pairs $\left(\frac{\pi}{2}, \text{undefined}\right)$, $\left(\frac{5\pi}{8}, 4\right)$, $\left(\frac{3\pi}{4}, 2\right)$, $\left(\frac{7\pi}{8}, 0\right)$, and $(\pi, \text{undefined})$ to sketch one cycle of $y = 2 \cot 2\left(x - \frac{\pi}{2}\right) + 2$.

Figure 6-5.4

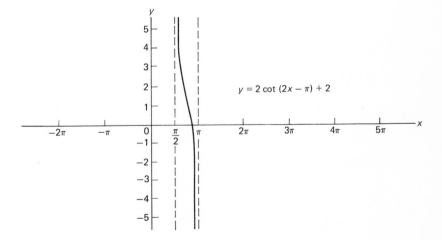

$y = 2 \cot (2x - \pi) + 2$

Exercise 6-5.1 In problems 1–5, which are of the form $y = a \cot (px + s) + h$, give the values for a, p, s, and h. Find the period, phase shift, and vertical shift of the graph.

1 $y = 3 \cot 2x$

2 $y = \frac{1}{3} \cot 3x$

3 $y = 2 \cot (2x - 3\pi) + 1$

4 $y = \frac{1}{2} \cot \left(x + \dfrac{\pi}{4}\right) - 2$

5 $y = -2 \cot (3x + 6) + \pi$

In problems 6–15, use the methods of this section to sketch one cycle of the given function.

6 $y = -\cot x$

7 $y = 5 \cot \left(x - \dfrac{\pi}{2}\right)$

8 $y = \cot (2x + \pi)$

9 $y = -2 \cot (x - \pi)$

10 $y = \cot \left(\frac{1}{2}x - \dfrac{\pi}{4}\right) + 1$

11 $y = -3 \cot \left(2x + \dfrac{\pi}{6}\right) + 2$

12 $y = \cot (3x + 4) - 1$

13 $y = \frac{4}{5} \cot \left(4x - \dfrac{\pi}{2}\right) + 2$

14 $y = 2 \cot \left(x + \dfrac{\pi}{6}\right) - 1$

15 $y = 3 \cot (3x - 2\pi) + \frac{1}{2}$

6-6 Graphing Secant Functions

To analyze the secant function more closely, we again consider a table of values and the graph of $y = \sec x$ for $-90° < x < 270°$. The graph of $y = \sec x$ is shown in Figure 6-6.1.

Table 6-6.1

θ	$-90°$	$-60°$	$-45°$	$-30°$	$0°$	$30°$	$45°$	$60°$	$90°$
$\sec \theta$	Undef.	2	1.4	1.2	1	1.2	1.4	2	Undef.

θ	$120°$	$135°$	$150°$	$180°$	$210°$	$225°$	$240°$	$270°$
$\sec \theta$	-2	-1.4	-1.2	-1	-1.2	-1.4	-2	Undef.

Figure 6-6.1

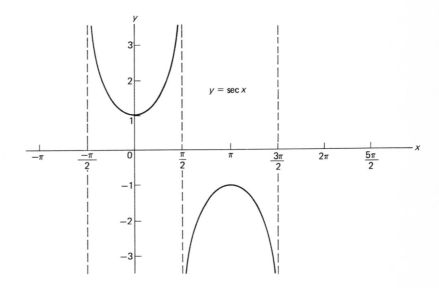

The period of $y = \sec x$ is 2π. The function is undefined when $x = -\dfrac{\pi}{2}$, $\dfrac{\pi}{2}$, and $\dfrac{3\pi}{2}$. It does not cross the x axis. The minimum positive value is 1 while the maximum negative value is -1. The y intercept occurs at $(0,1)$. Table 6-6.2 gives the reference points for $y = \sec x$.

Table 6-6.2
Reference points for $y = \sec x$

x	$-\dfrac{\pi}{2}$	0	$\dfrac{\pi}{2}$	π	$\dfrac{3\pi}{2}$
y	Undef.	1	Undef.	-1	Undef.

We now apply our method of rapid sketching to sketch the graphs of equations of the form $y = a \sec (px + s) + h$.

Study the following examples.

Example 6-6.1 Sketch one cycle of the graph of $y = 2 \sec 2x$.

Solution **STEP 1** Factor out p. Unnecessary.

STEP 2 Find the period.

$$\frac{1}{|2|} \cdot 2\pi = \pi$$

STEP 3 Construct a reference table as shown in Table 6-6.3.

Table 6-6.3
Reference table for $y = 2 \sec 2x$

x	$-\dfrac{\pi}{4}$	0	$\dfrac{\pi}{4}$	$\dfrac{\pi}{2}$	$\dfrac{3\pi}{4}$	
$2x$	$-\dfrac{\pi}{2}$	0	$\dfrac{\pi}{2}$	π	$\dfrac{3\pi}{2}$	← Fill in first
$\sec 2x$	Undef.	1	Undef.	-1	Undef.	
$y = 2 \sec 2x$	Undef.	2	Undef.	-2	Undef.	

STEP 4 Use the ordered pairs $\left(-\dfrac{\pi}{4}, \text{undefined}\right)$, $(0,2)$, $\left(\dfrac{\pi}{4}, \text{undefined}\right)$, $\left(\dfrac{\pi}{4}, -2\right)$, and $(\tfrac{3}{4}, \text{undefined})$ to sketch one cycle of $y = 2 \sec 2x$.

Figure 6-6.2

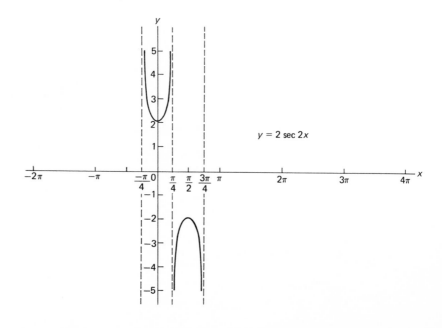

$y = 2 \sec 2x$

Example 6-6.2   Sketch one cycle of $y = \frac{1}{8} \sec \left(\frac{1}{2}x + \pi\right)$.

Solution   **STEP 1**   Factor out $\frac{1}{2}$.

$$y = \tfrac{1}{8} \sec \tfrac{1}{2}(x + 2\pi)$$

STEP 2   Find the period.

$$\frac{1}{\left|\frac{1}{2}\right|} \cdot 2\pi = 4\pi$$

STEP 3   Construct a reference table for $y = \frac{1}{8} \sec \frac{1}{2}(x + 2\pi)$.

Table 6-6.4

Reference table for
$y = \frac{1}{8} \sec \frac{1}{2}(x + 2\pi)$

x	-3π	-2π	$-\pi$	0	π	
$x + 2\pi$	$-\pi$	0	π	2π	3π	
$\frac{1}{2}(x + 2\pi)$	$-\dfrac{\pi}{2}$	0	$\dfrac{\pi}{2}$	π	$\dfrac{3\pi}{2}$	← Fill in first
$\sec \frac{1}{2}(x + 2\pi)$	Undef.	1	Undef.	-1	Undef.	
$y = \frac{1}{8} \sec \frac{1}{2}(x + 2\pi)$	Undef.	$\dfrac{1}{8}$	Undef.	$-\dfrac{1}{8}$	Undef.	

STEP 4   Sketch one cycle of $y = \frac{1}{8} \sec \frac{1}{2}(x + \pi)$ using the ordered pairs $(-3\pi, \text{undefined})$, $(-2\pi, \frac{1}{8})$, $(-\pi, \text{undefined})$, $(0, -\frac{1}{8})$, and $(\pi, \text{undefined})$.

Figure 6-6.3

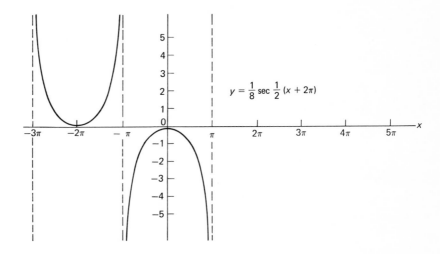

$$y = \frac{1}{8} \sec \frac{1}{2}(x + 2\pi)$$

Example 6-6.3   Sketch one cycle of $y = 3 \sec \left(2x - \dfrac{\pi}{2}\right) - 1$.

Solution   **STEP 1**   Factor out 2.

$$y = 3 \sec 2\left(x - \frac{\pi}{4}\right) - 1$$

STEP 2 Find the period.

$$\frac{1}{|2|} \cdot 2\pi = \pi$$

STEP 3 Construct a reference table for $y = 3 \sec 2\left(x - \frac{\pi}{4}\right) - 1$.

Table 6-6.5

Reference table for $y = 3 \sec 2\left(x - \frac{\pi}{4}\right) - 1$

x	0	$\frac{\pi}{4}$	$\frac{\pi}{2}$	$\frac{3\pi}{4}$	π	
$x - \frac{\pi}{4}$		$-\frac{\pi}{4}$	0	$\frac{\pi}{4}$	$\frac{\pi}{2}$	$\frac{3\pi}{4}$
$2\left(x - \frac{\pi}{4}\right)$		$-\frac{\pi}{2}$	0	$\frac{\pi}{2}$	π	$\frac{3\pi}{2}$ ← Fill in first
$\sec 2\left(x - \frac{\pi}{4}\right)$		Undef.	1	Undef.	−1	Undef.
$3 \sec 2\left(x - \frac{\pi}{4}\right)$		Undef.	3	Undef.	−3	Undef.
$y = 3 \sec 2\left(x - \frac{\pi}{4}\right) - 1$		Undef.	2	Undef.	−4	Undef.

STEP 4 Sketch one cycle of $y = 3 \sec \left(2x - \frac{\pi}{2}\right) - 1$ using the ordered pairs $(0, \text{undefined})$, $\left(\frac{\pi}{4}, 2\right)$, $\left(\frac{\pi}{2}, \text{undefined}\right)$, $\left(\frac{3\pi}{4}, -4\right)$, and $(\pi, \text{undefined})$.

Figure 6-6.4

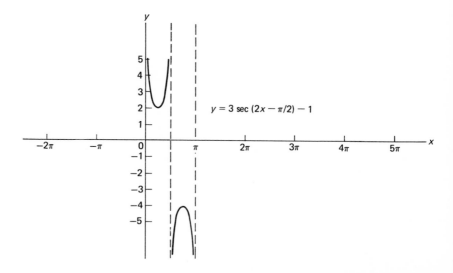

$y = 3 \sec (2x - \pi/2) - 1$

Exercise 6-6.1 In problems 1-5, which are of the form $y = a \sec (px + s) + h$, give the values of a, p, s, and h. Find the positive minimum, negative maximum, period, phase shift, and vertical shift.

1 $y = 3 \sec \frac{1}{2}x$

2 $y = 2 \sec (3x + \pi)$

3 $y = -\frac{1}{2} \sec \left(2x - \frac{\pi}{2}\right)$

4 $y = 3 \sec (\frac{1}{2}x + \pi) - 1$

5 $y = 4 \sec \left(2x - \frac{\pi}{4}\right) + 2$

Use the methods of this section to sketch one cycle of the graph of each of the following:

6 $y = \sec 3x$

7 $y = -\sec \frac{1}{3}x$

8 $y = 2 \sec \frac{\pi}{4}x$

9 $y = 3 \sec (4x + \pi)$

10 $y = -2 \sec \left(\frac{1}{2}x - \frac{\pi}{4}\right)$

11 $y = 2 \sec \left(3x + \frac{\pi}{2}\right)$

12 $y = \sqrt{7} \sec (-x + \pi) + 1$

13 $y = \frac{1}{2} \sec \left(\frac{1}{2}x - \frac{\pi}{4}\right) - 2$

14 $y = \frac{1}{3} \sec (x - 3) + 1$

15 $y = 2 \sec \left(\frac{x}{3} + \frac{\pi}{2}\right) - 2$

6-7 Graphing Cosecant Functions

To examine the variation of the cosecant function and some of the properties of $y = \csc x$, we make the following table of values (Table 6-7.1) and a graph of $y = \csc x$ (Figure 6-7.1).

Table 6-7.1

θ	0°	30°	45°	60°	90°	120°	135°	150°	180°
$\csc \theta$	Undef.	2	1.4	1.2	1	1.2	1.4	2	Undef.

θ	210°	225°	240°	270°	300°	315°	330°	360°
$\csc \theta$	−2	−1.4	−1.2	−1	−1.2	−1.4	−2	Undef.

Figure 6-7.1

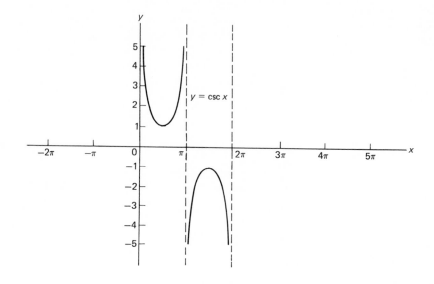

Figure 6-7.1 shows that the shape of the curve $y = \csc x$ is identical to the curve $y = \sec x$. It has the same smallest positive value, 1, and the same largest negative value, −1. The period is 2π, and the function is undefined when $x = 0$, π, and 2π.

Table 6-7.2 summarizes the reference points we can use for sketching graphs that involve the cosecant function.

Table 6-7.2
Reference table for $y = \csc x$

x	0	$\dfrac{\pi}{2}$	π	$\dfrac{3\pi}{2}$	2π
$y = \csc x$	Undef.	1	Undef.	−1	Undef.

We now use our familiar method of rapid sketching to sketch one cycle of the graphs of equations of the form $y = a \csc (px + s) + h$.

The following examples illustrate the technique. Remember that a

affects the amplitude, p affects the period, $\dfrac{s}{p}$ affects the horizontal shift, and h affects the vertical shift.

Example 6-7.1 Sketch one cycle of $y = 2 \csc 2x$.

Solution **STEP 1** Factor out p. Unnecessary.

STEP 2 Find the period.

$$\frac{1}{|2|} \cdot 2\pi = \pi$$

STEP 3 Construct a reference table for $y = 2 \csc 2x$.

Table 6-7.3
Reference table for
$y = 2 \csc 2x$

x	0	$\dfrac{\pi}{4}$	$\dfrac{\pi}{2}$	$\dfrac{3\pi}{4}$	π	
$2x$	0	$\dfrac{\pi}{2}$	π	$\dfrac{3\pi}{2}$	2π	← Fill in first
$\csc 2x$	Undef.	1	Undef.	-1	Undef.	
$y = 2 \csc 2x$	Undef.	2	Undef.	-2	Undef.	

STEP 4 Sketch one cycle of $y = 2 \csc 2x$ using the ordered pairs $(0, \text{undefined})$, $\left(\dfrac{\pi}{4}, 2\right)$, $\left(\dfrac{\pi}{2}, \text{undefined}\right)$, $\left(\dfrac{3\pi}{4}, -2\right)$, and $(\pi, \text{undefined})$ from the reference table.

Figure 6-7.2

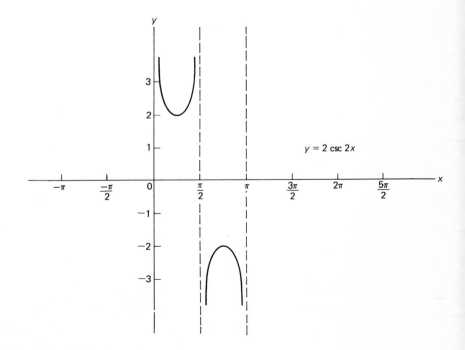

Example 6-7.2 Sketch one cycle of the graph of $y = \frac{3}{2} \csc \left(\frac{x}{2} - \pi \right)$.

Solution **STEP 1** Factor out $\frac{1}{2}$:
$$y = \tfrac{3}{2} \csc \tfrac{1}{2}(x - 2\pi)$$

STEP 2 Find the period:

$$\frac{1}{\left| \frac{1}{2} \right|} \cdot 2\pi = 4\pi$$

STEP 3 Construct a reference table for $y = \frac{3}{2} \csc \frac{1}{2}(x - 2\pi)$.

Table 6-7.4

Reference table for $y = \frac{3}{2} \csc \frac{1}{2}(x - 2\pi)$

x	2π	3π	4π	5π	6π	
$x - 2\pi$	0	π	2π	3π	4π	
$\frac{1}{2}(x - 2\pi)$	0	$\frac{\pi}{2}$	π	$\frac{3\pi}{2}$	2π	←Fill in first
$\csc \frac{1}{2}(x - 2\pi)$	Undef.	1	Undef.	-1	Undef.	
$y = \frac{3}{2} \csc \frac{1}{2}(x - 2\pi)$	Undef.	$\frac{3}{2}$	Undef.	$-\frac{3}{2}$	Undef.	

STEP 4 Sketch one cycle of $y = \frac{3}{2} \csc \left(\frac{x}{2} - \pi \right)$ using the ordered pairs $(2\pi, \text{undefined})$, $(3\pi, \frac{3}{2})$, $(4\pi, \text{undefined})$, $(5\pi, -\frac{3}{2})$, and $(6\pi, \text{undefined})$.

Figure 6-7.3

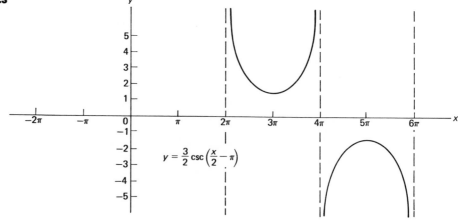

$$y = \frac{3}{2} \csc \left(\frac{x}{2} - \pi \right)$$

Example 6-7.3   Sketch one cycle of the graph of $y = 2 \csc (2x + \pi) - 2$.

Solution   **STEP 1**   Factor out 2.

$$y = 2 \csc 2\left(x + \frac{\pi}{2}\right) - 2$$

STEP 2   Find the period.

$$\frac{1}{|2|} \cdot 2\pi = \pi$$

STEP 3   Construct a reference table for $y = 2 \csc 2\left(x + \frac{\pi}{2}\right) - 2$.

Table 6-7.5

Reference table for
$y = 2 \csc 2\left(x + \frac{\pi}{2}\right) - 2$

x	$-\dfrac{\pi}{2}$	$-\dfrac{\pi}{4}$	0	$\dfrac{\pi}{4}$	$\dfrac{\pi}{2}$	
$x + \dfrac{\pi}{2}$	0	$\dfrac{\pi}{4}$	$\dfrac{\pi}{2}$	$\dfrac{3\pi}{4}$	π	
$2\left(x + \dfrac{\pi}{2}\right)$	0	$\dfrac{\pi}{2}$	π	$\dfrac{3\pi}{2}$	2π	← Fill in first
$\csc 2\left(x + \dfrac{\pi}{2}\right)$	Undef.	1	Undef.	-1	Undef.	
$2 \csc 2\left(x + \dfrac{\pi}{2}\right)$	Undef.	2	Undef.	-2	Undef.	
$y = 2 \csc 2\left(x + \dfrac{\pi}{2}\right) - 2$	Undef.	0	Undef.	-4	Undef.	

STEP 4   Sketch one cycle of the graph using the ordered pairs $\left(-\dfrac{\pi}{2}, \text{unde-}\right.$ fined$\Big)$, $\left(-\dfrac{\pi}{4}, 0\right)$, $(0, \text{undefined})$, $\left(\dfrac{\pi}{4}, -4\right)$, and $\left(\dfrac{\pi}{2}, \text{undefined}\right)$ given in the reference table.

Figure 6-7.4

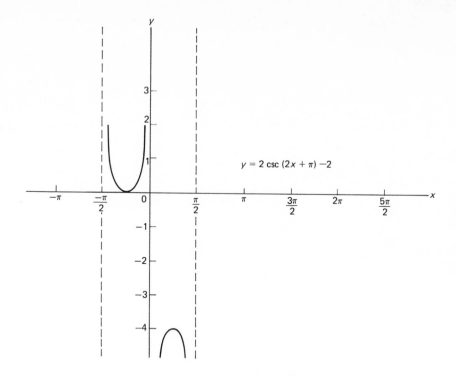

$y = 2 \csc (2x + \pi) - 2$

Exercise 6-7.1 In problems 1–5, which are of the form $y = a \csc (px + s) + h$, give the values of a, p, s, and h and find the period, phase shift, and vertical shift.

1 $y = 4 \csc \tfrac{1}{2}x$

2 $y = \csc (2x - \pi)$

3 $y = -2 \csc \left(3x + \dfrac{\pi}{2}\right)$

4 $y = 5 \csc \left(\dfrac{x}{2} - 3\pi\right) - 1$

5 $y = -\tfrac{4}{5} \csc \left(\dfrac{x}{4} - \dfrac{3\pi}{2}\right) + 3$

In problems 6–15, use the methods of this chapter to sketch one cycle of the graph.

6 $y = \csc \tfrac{1}{8}x$

7 $y = 8 \csc x$

8 $y = 2 \csc \frac{2}{3}x$

9 $y = \csc (\pi x + 1)$

10 $y = \csc (3x + \pi)$

11 $y = \sqrt{3} \csc (x + \pi) - 2$

12 $y = -2 \csc \left(\dfrac{2x + \pi}{3}\right) - 2$

13 $y = 3 \csc (x - 2\pi) - 1$

14 $y = -\frac{1}{2} \csc (2x + 3) + 1$

15 $y = 2 \csc (\pi - x) + 3$

Review Exercise Graph one cycle for each of the following:

1 $y = \sqrt{3} \sin \frac{1}{4}x$

2 $y = \tan \left(2x - \dfrac{\pi}{6}\right)$

3 $y = -\csc 7x$

4 $y = 5 \tan 2\pi x$

5 $y = \cos (3x - 1)$

6 $y = k \sec \frac{1}{9}x, \; k < 0$

7 $y = \frac{2}{3} \cot \left(x + \dfrac{\pi}{8}\right)$

8 $y = \tan (x + 2)$

9 $y = -\cos 5x$

10 $y = \frac{1}{7} \sin (\frac{1}{5}x - \pi)$

11 $y = k \cos \frac{1}{3}x, \ k > 0$

12 $y = k \sin \frac{1}{3}x, \ k > 0$

Answers

Exercise 6-2.1

1 $y = 2 \sin x; \ a = 2; \ p = 1; \ s = 0; \ h = 0;$ amplitude $= 2;$ period $= 2\pi;$ phase shift $= 0;$ vertical shift $= 0.$

2 $y = -\frac{1}{3} \sin (2x + \pi); \ a = -\frac{1}{3}; \ p = 2; \ s = \pi; \ h = 0;$ amplitude $= \frac{1}{3};$ period $= \pi;$ phase shift $= \dfrac{\pi}{2};$ vertical shift $= 0.$

3 $y = \frac{1}{5} \sin (\frac{1}{2}x + 2\pi) + 4; \ a = \frac{1}{5}; \ p = \frac{1}{2}; \ s = 2\pi; \ h = 4;$ amplitude $= \frac{1}{5};$ period $= 4\pi;$ phase shift $= 4\pi;$ vertical shift $= 4.$

4 $y = 5 \sin (-2x + \pi) - 2; \ a = 5; \ p = -2; \ s = \pi; \ h = -2;$ amplitude $= 5;$ period $= \pi;$ phase shift $= -\dfrac{\pi}{2};$ vertical shift $= -2.$

5 $y = -2 \sin (\frac{1}{3}x + 3\pi) + \pi; \ a = -2; \ p = \frac{1}{3}; \ s = 3\pi; \ h = \pi;$ amplitude $= 2;$ period $= 6\pi;$ phase shift $= 9\pi;$ vertical shift $= \pi.$

6 Reference points: $(0,0),$ $\left(\dfrac{\pi}{2},3\right),$ $(\pi,0),$ $\left(\dfrac{3\pi}{2},-3\right),$ $(2\pi,0)$

7 Reference points: $(0,0),$ $(\pi,2),$ $(2\pi,0),$ $(3\pi,-2),$ $(4\pi,0)$

8 Reference points: $(0,0),$ $\left(\dfrac{\pi}{4},\dfrac{1}{2}\right),$ $\left(\dfrac{\pi}{2},0\right),$ $\left(\dfrac{3\pi}{4},-\dfrac{1}{2}\right),$ $(\pi,0)$

9 Reference points: $\left(-\dfrac{\pi}{3},0\right),$ $\left(-\dfrac{\pi}{6},\dfrac{1}{8}\right),$ $(0,0),$ $\left(\dfrac{\pi}{6},-\dfrac{1}{8}\right),$ $\left(\dfrac{\pi}{3},0\right)$

10 Reference points: $(0,0),$ $\left(\dfrac{\pi}{8},-3\right),$ $\left(\dfrac{\pi}{4},0\right),$ $\left(\dfrac{3\pi}{2},3\right),$ $\left(\dfrac{\pi}{2},0\right)$

11 Reference points: $\left(-\dfrac{\pi}{3},0\right),$ $\left(-\dfrac{\pi}{6},1\right),$ $(0,0),$ $\left(\dfrac{\pi}{6},-1\right),$ $\left(\dfrac{\pi}{3},0\right)$

12 Reference points: $(0,2)$, $\left(\dfrac{3\pi}{4},3\right)$, $\left(\dfrac{3\pi}{2},2\right)$, $\left(\dfrac{9\pi}{4},1\right)$, $(3\pi,2)$

13 Reference points: $(-3\pi,1)$, $\left(-\dfrac{3\pi}{2},3\right)$, $(0,1)$, $\left(\dfrac{3\pi}{2},-1\right)$, $(3\pi,1)$

14 Reference points: $(-2\pi,2)$, $\left(-\dfrac{3\pi}{2},1\right)$, $(-\pi,2)$, $\left(-\dfrac{\pi}{2},3\right)$, $(0,2)$

15 Reference points: $(-\pi,-1)$, $\left(-\dfrac{3\pi}{4},-\dfrac{1}{2}\right)$, $\left(-\dfrac{\pi}{2},-1\right)$, $\left(-\dfrac{\pi}{4},-\dfrac{3}{2}\right)$, $(0,-1)$

Exercise 6-3.1

1 $y = \cos 3x$; $a = 1$; $p = 3$; $s = 0$; $h = 0$; amplitude $= 1$; period $= \dfrac{2\pi}{3}$; phase shift $= 0$; vertical shift $= 0$.

2 $a = 2$; $p = 4$; $s = 0$; $h = 0$; amplitude $= 2$; period $= \dfrac{\pi}{2}$; phase shift $= 0$; vertical shift $= 0$.

3 $y = \frac{1}{2}\cos(x + \pi)$; $a = \frac{1}{2}$; $p = 1$; $s = \pi$; $h = 0$; amplitude $= \frac{1}{2}$; period $= 2\pi$; phase shift $= \pi$; vertical shift $= 0$.

4 $a = -2$; $p = 2$; $s = \dfrac{\pi}{2}$; $h = 0$; amplitude $= 2$; period $= \pi$; phase shift $= \dfrac{\pi}{4}$; vertical shift $= 0$.

5 $y = 3\cos(3x - \pi) + 2$; $a = 3$; $p = 3$; $s = -\pi$; $h = 2$; amplitude $= 3$; period $= \dfrac{2\pi}{3}$; phase shift $= -\dfrac{\pi}{3}$; vertical shift $= 2$.

6 Reference points: $(0,3)$, $\left(\dfrac{\pi}{2},0\right)$, $(\pi,-3)$, $\left(\dfrac{3\pi}{2},0\right)$, $(2\pi,3)$

7 Reference points: $\left(-\dfrac{\pi}{2},1\right)$, $\left(-\dfrac{\pi}{4},0\right)$, $(0,-1)$, $\left(\dfrac{\pi}{4},0\right)$, $\left(\dfrac{\pi}{2},1\right)$

8 Reference points: $\left(\dfrac{\pi}{2},\dfrac{1}{2}\right)$, $\left(\dfrac{3\pi}{4},0\right)$, $\left(\pi,-\dfrac{1}{2}\right)$, $\left(\dfrac{5\pi}{4},0\right)$, $\left(\dfrac{3\pi}{2},\dfrac{1}{2}\right)$

9 Reference points: $(-4\pi,-2)$, $(-3\pi,0)$, $(-2\pi,2)$, $(-\pi,0)$, $(0,-2)$

10 Reference points: $\left(-\dfrac{\pi}{2},\dfrac{3}{2}\right)$, $(0,0)$, $\left(\dfrac{\pi}{2},-\dfrac{3}{2}\right)$, $(\pi,0)$, $\left(\dfrac{3\pi}{2},\dfrac{3}{2}\right)$

11 Reference points: $\left(-\dfrac{1}{3},-3\right)$, $\left(\dfrac{\pi}{6}-\dfrac{1}{3},0\right)$, $\left(\dfrac{\pi}{3}-\dfrac{1}{3},3\right)$, $\left(\dfrac{\pi}{2}-\dfrac{1}{3},0\right)$, $\left(\dfrac{2\pi}{3}-\dfrac{1}{3},-3\right)$

12 Reference points: $(-2\pi, 3)$, $\left(-\dfrac{3\pi}{2}, 1\right)$, $(-\pi, -1)$, $\left(-\dfrac{\pi}{2}, 1\right)$, $(0, 3)$

13 Reference points: $\left(\dfrac{\pi}{2}, 1\right)$, $\left(\dfrac{5\pi}{8}, -2\right)$, $\left(\dfrac{3\pi}{4}, -5\right)$, $\left(\dfrac{7\pi}{8}, -2\right)$, $(\pi, 1)$

14 Reference points: $\left(-2\pi, \dfrac{1}{4}\right)$, $(-\pi, 1)$, $\left(0, \dfrac{7}{4}\right)$, $(\pi, 1)$, $\left(2\pi, \dfrac{1}{4}\right)$

15 Reference points: $\left(\dfrac{\pi}{4}, 3\right)$, $(0, -1)$, $\left(-\dfrac{\pi}{4}, -5\right)$, $\left(-\dfrac{\pi}{2}, -1\right)$, $\left(-\dfrac{3\pi}{4}, 3\right)$

Exercise 6-4.1

1 $y = 3 \tan (2x + 3) + 5$

$a = 3$; $p = 2$; $s = 3$; $h = 5$; period $= \dfrac{\pi}{2}$; phase shift $= \tfrac{3}{2}$; vertical shift $= 5$

2 $y = \tan (x - 4) - 1$
$a = 1$; $p = 1$; $s = -4$; $h = -1$; period $= \pi$; phase shift $=$ right 4 units; vertical shift $= -1$

3 $y = \tfrac{1}{2} \tan (2x - \pi)$

$a = \tfrac{1}{2}$; $p = 2$; $s = -\pi$; $h = 0$; period $= \dfrac{\pi}{2}$; phase shift $= \dfrac{\pi}{2}$ right; vertical shift $= 0$

4 $y = 5 \tan \left(\tfrac{1}{3}x - \dfrac{\pi}{2}\right) + 1$

$a = 5$; $p = \dfrac{1}{3}$; $s = -\dfrac{\pi}{2}$; $h = 1$; period $= 3\pi$; phase shift $= \dfrac{3\pi}{2}$; vertical shift $= 1$

5 $y = 6 \tan (\tfrac{1}{5}x - \pi) - 2$
$a = 6$; $p = \tfrac{1}{5}$; $s = -\pi$; $h = -2$; period $= 5\pi$; phase shift $= 5\pi$ right; vertical shift $= -2$

6 Reference points: $\left(\dfrac{\pi}{6}, \text{undef.}\right)$, $\left(\dfrac{\pi}{12}, 2\right)$, $(0, 0)$, $\left(-\dfrac{\pi}{6}, \text{undef.}\right)$, $\left(-\dfrac{\pi}{12}, -2\right)$

7 Reference points: $(-2\pi, \text{undef.})$, $\left(-\pi, -\dfrac{1}{2}\right)$, $(0, 0)$, $\left(\pi, \dfrac{1}{2}\right)$, $(2\pi, \text{undef.})$

8 Reference points: $\left(-\dfrac{\pi}{12}, \text{undef.}\right)$, $\left(-\dfrac{\pi}{24}, \dfrac{1}{4}\right)$, $(0, 0)$, $\left(\dfrac{\pi}{24}, -\dfrac{1}{4}\right)$, $\left(\dfrac{\pi}{12}, \text{undef.}\right)$

9 Reference points: $\left(\dfrac{\pi}{6}, \text{undef.}\right)$, $\left(\dfrac{\pi}{4}, -3\right)$, $\left(\dfrac{\pi}{3}, 0\right)$, $\left(\dfrac{5\pi}{12}, 3\right)$, $\left(\dfrac{\pi}{2}, \text{undef.}\right)$

10 Reference points: $\left(1 - \dfrac{\pi}{4}, \text{undef.}\right)$, $\left(1 - \dfrac{\pi}{8}, -\dfrac{1}{2}\right)$, $(1,0)$, $\left(1 + \dfrac{\pi}{8}, \dfrac{1}{2}\right)$ $\left(1 + \dfrac{\pi}{4}, \text{undef.}\right)$

11 Reference points: $\left(-\dfrac{5\pi}{8}, \text{undef.}\right)$, $\left(-\dfrac{9\pi}{16}, -2\right)$, $\left(-\dfrac{\pi}{2}, 0\right)$, $\left(-\dfrac{7\pi}{16}, 2\right)$, $\left(-\dfrac{3\pi}{8}, \text{undef.}\right)$

12 Reference points: $(0, \text{undef.})$, $\left(\dfrac{\pi}{8}, 0\right)$, $\left(\dfrac{\pi}{4}, 1\right)$, $\left(\dfrac{3\pi}{8}, 2\right)$, $\left(\dfrac{\pi}{2}, \text{undef.}\right)$

13 Reference points: $\left(\dfrac{\pi}{2}, \text{undef.}\right)$, $\left(\dfrac{3\pi}{4}, -5\right)$, $(\pi, -3)$, $\left(\dfrac{5\pi}{4}, -1\right)$, $\left(\dfrac{3\pi}{2}, \text{undef.}\right)$

14 Reference points: $(-6\pi, \text{undef.})$, $(-5\pi, 3)$, $(-4\pi, 1)$, $(-3\pi, -1)$, $(-2\pi, \text{undef.})$

15 Reference points: $\left(-\dfrac{3\pi}{8}, \text{undef.}\right)$, $\left(-\dfrac{\pi}{4}, -5\right)$, $\left(-\dfrac{\pi}{8}, -2\right)$, $(0,1)$, $\left(\dfrac{\pi}{8}, \text{undef.}\right)$

Exercise 6-5.1

1 $a = 3$; $p = 2$; $s = 0$; $h = 0$; period $= \dfrac{\pi}{2}$; phase shift $= 0$; vertical shift $= 0$

2 $a = \dfrac{1}{3}$; $p = 3$; $s = 0$; $h = 0$; period $= \dfrac{\pi}{3}$; phase shift $= 0$; vertical shift $= 0$

3 $a = 2$; $p = 2$; $s = -3\pi$; $h = 1$; period $= \dfrac{\pi}{2}$; phase shift $= \dfrac{3\pi}{2}$; vertical shift $= 1$

4 $a = \dfrac{1}{2}$; $p = 1$; $s = \dfrac{\pi}{4}$; $h = -2$; period $= \pi$; phase shift $= \dfrac{\pi}{4}$; vertical shift $= -2$

5 $a = -2$; $p = 3$; $s = 6$; $h = \pi$; period $= \dfrac{\pi}{3}$; phase shift $= 2$; vertical shift $= \pi$

6 Reference points: $(0, \text{undef.})$, $\left(\frac{\pi}{4}, -1\right)$, $\left(\frac{\pi}{2}, 0\right)$, $\left(\frac{3\pi}{4}, 1\right)$, $(\pi, \text{undef.})$

7 Reference points: $\left(\frac{\pi}{2}, \text{undef.}\right)$, $\left(\frac{3\pi}{4}, 5\right)$, $(\pi, 0)$, $\left(\frac{5\pi}{4}, -5\right)$, $\left(\frac{3\pi}{2}, \text{undef.}\right)$

8 Reference points: $\left(-\frac{\pi}{2}, \text{undef.}\right)$, $\left(-\frac{3\pi}{8}, 1\right)$, $\left(-\frac{\pi}{4}, 0\right)$, $\left(-\frac{\pi}{8}, -1\right)$, $(0, \text{undef.})$

9 Reference points: $(\pi, \text{undef.})$, $\left(\frac{5\pi}{4}, -2\right)$, $\left(\frac{3\pi}{2}, 0\right)$, $\left(\frac{7\pi}{4}, 2\right)$, $(2\pi, \text{undef.})$

10 Reference points: $\left(\frac{\pi}{2}, \text{undef.}\right)$, $(\pi, 2)$, $\left(\frac{3\pi}{2}, 1\right)$, $(2\pi, 0)$, $\left(\frac{5\pi}{2}, \text{undef.}\right)$

11 Reference points: $\left(-\frac{\pi}{12}, \text{undef.}\right)$, $\left(\frac{\pi}{24}, -1\right)$, $\left(\frac{\pi}{6}, 2\right)$, $\left(\frac{7\pi}{24}, 5\right)$, $\left(\frac{5\pi}{12}, \text{undef.}\right)$

12 Reference points: $\left(-\frac{4}{3}, \text{undef.}\right)$, $\left(\frac{\pi}{12} - \frac{4}{3}, 0\right)$, $\left(\frac{\pi}{6} - \frac{4}{3}, -1\right)$, $\left(\frac{\pi}{4} - \frac{4}{3}, -2\right)$, $\left(\frac{\pi}{3} - \frac{4}{3}, \text{undef.}\right)$

13 Reference points: $\left(\frac{\pi}{8}, \text{undef.}\right)$, $\left(\frac{3\pi}{16}, \frac{14}{5}\right)$, $\left(\frac{\pi}{4}, 2\right)$, $\left(\frac{5\pi}{16}, \frac{6}{5}\right)$, $\left(\frac{3\pi}{8}, \text{undef.}\right)$

14 Reference points: $\left(-\frac{\pi}{6}, \text{undef.}\right)$, $\left(\frac{\pi}{12}, 1\right)$, $\left(\frac{\pi}{3}, -1\right)$, $\left(\frac{7\pi}{12}, -3\right)$, $\left(\frac{5\pi}{6}, \text{undef.}\right)$

15 Reference points: $\left(\frac{2\pi}{3}, \text{undef.}\right)$, $\left(\frac{3\pi}{4}, \frac{7}{2}\right)$, $\left(\frac{5\pi}{6}, \frac{1}{2}\right)$, $\left(\frac{11\pi}{12}, -\frac{5}{2}\right)$, $(\pi, \text{undef.})$

Exercise 6-6.1

1 $y = 3 \sec \frac{1}{2}x$
  $a = 3; p = \frac{1}{2}; s = 0; h = 0$; positive minimum $= 3$; negative maximum $= -3$; period $= 4\pi$; phase shift $= 0$; vertical shift $= 0$

2 $y = 2 \sec (3x + \pi)$
  $a = 2; p = 3; s = \pi; h = 0$; positive minimum $= 2$; negative maximum $= -2$; period $= \frac{2\pi}{3}$; phase shift $= \frac{\pi}{3}$; vertical shift $= 0$

3 $y = -\frac{1}{2} \sec \left(2x - \frac{\pi}{2} \right)$

$a = -\frac{1}{2}$; $p = 2$; $s = -\frac{\pi}{2}$; $h = 0$; positive minimum $= \frac{1}{2}$; negative

maximum $= -\frac{1}{2}$; period $= \pi$; phase shift $= -\frac{\pi}{4}$; vertical shift $= 0$

4 $y = 3 \sec \left(\frac{1}{2}x + \pi \right) - 1$
$a = 3$; $p = \frac{1}{2}$; $s = \pi$; $h = -1$; positive minimum $= 2$; negative maximum $= -4$; period $= 4\pi$; phase shift $= 2\pi$; vertical shift $= -1$

5 $y = 4 \sec \left(2x - \frac{\pi}{4} \right) + 2$

$a = 4$; $p = 2$; $s = -\frac{\pi}{4}$; $h = 2$; positive minimum $= 6$; negative

maximum $= -2$; period $= \pi$; phase shift $= -\frac{\pi}{8}$; vertical shift $= 2$

6 Reference points: $\left(-\frac{\pi}{6}, \text{undef.} \right)$, $(0,1)$, $\left(\frac{\pi}{6}, \text{undef.} \right)$, $\left(\frac{\pi}{3}, -1 \right)$,

$\left(\frac{\pi}{2}, \text{undef.} \right)$

7 Reference points: $\left(-\frac{3\pi}{2}, \text{undef.} \right)$, $(0,-1)$, $\left(\frac{3\pi}{2}, \text{undef.} \right)$, $(3\pi, 1)$,

$\left(\frac{9\pi}{2}, \text{undef.} \right)$

8 Reference points: $(-2, \text{undef.})$, $(0,2)$, $(2, \text{undef.})$, $(4,-2)$, $(6, \text{undef.})$

9 Reference points: $\left(-\frac{3\pi}{8}, \text{undef.} \right)$, $\left(-\frac{\pi}{4}, 3 \right)$, $\left(-\frac{\pi}{8}, \text{undef.} \right)$, $(0,-3)$,

$\left(\frac{\pi}{8}, \text{undef.} \right)$

10 Reference points: $\left(-\frac{\pi}{2}, \text{undef.} \right)$, $\left(\frac{\pi}{2}, -2 \right)$, $\left(\frac{3\pi}{2}, \text{undef.} \right)$, $\left(\frac{5\pi}{2}, 2 \right)$,

$\left(\frac{7\pi}{2}, \text{undef.} \right)$

11 Reference points: $\left(-\frac{\pi}{3}, \text{undef.} \right)$, $\left(-\frac{\pi}{6}, 2 \right)$, $(0, \text{undef.})$, $\left(\frac{\pi}{6}, -2 \right)$,

$\left(\frac{\pi}{3}, \text{undef.} \right)$

12 Reference points: $\left(-\frac{\pi}{2}, \text{undef.} \right)$, $(0, 1 - \sqrt{7})$, $\left(\frac{\pi}{2}, \text{undef.} \right)$,

$(\pi, 1 + \sqrt{7})$, $\left(\frac{3\pi}{2}, \text{undef.} \right)$

13 Reference points: $\left(-\dfrac{\pi}{2}, \text{undef.}\right)$, $\left(\dfrac{\pi}{2}, -\dfrac{3}{2}\right)$, $\left(\dfrac{3\pi}{2}, \text{undef.}\right)$, $\left(\dfrac{5\pi}{2}, -\dfrac{5}{2}\right)$, $\left(\dfrac{7\pi}{2}, \text{undef.}\right)$

14 Reference points: $\left(3 - \dfrac{\pi}{2}, \text{undef.}\right)$, $(3, \frac{4}{3})$, $\left(3 + \dfrac{\pi}{2}, \text{undef.}\right)$, $(3 + \pi, \frac{2}{3})$, $\left(3 + \dfrac{3\pi}{2}, \text{undef.}\right)$

15 Reference points: $(-3\pi, \text{undef.})$, $\left(-\dfrac{3\pi}{2}, 0\right)$, $(0, \text{undef.})$, $\left(\dfrac{3\pi}{2}, -4\right)$, $(3\pi, \text{undef.})$

Exercise 6-7.1

1 $a = 4$; $p = \frac{1}{2}$; $s = 0$; $h = 0$; period $= 4\pi$; phase shift $= 0$; vertical shift $= 0$

2 $a = 1$; $p = 2$; $s = -\pi$; $h = 0$; period $= \pi$; phase shift $= -\dfrac{\pi}{2}$; vertical shift $= 0$

3 $a = -2$; $p = 3$; $s = \dfrac{\pi}{2}$; $h = 0$; period $= \dfrac{2\pi}{3}$; phase shift $= \dfrac{\pi}{6}$; vertical shift $= 0$

4 $a = 5$; $p = \frac{1}{2}$; $s = -3\pi$; $h = -1$; period $= 4\pi$; phase shift $= -6\pi$; vertical shift $= -1$

5 $a = -\dfrac{4}{5}$; $p = \dfrac{1}{4}$; $s = -\dfrac{3\pi}{2}$; $h = 3$; period $= 8\pi$; phase shift $= -6\pi$; vertical shift $= 3$

6 Reference points: $(0, \text{undef.})$, $(4\pi, 1)$, $(8\pi, \text{undef.})$, $(12\pi, -1)$, $(16\pi, \text{undef.})$

7 Reference points: $(0, \text{undef.})$, $\left(\dfrac{\pi}{2}, 8\right)$, $(\pi, \text{undef.})$, $\left(\dfrac{3\pi}{2}, -8\right)$, $(2\pi, \text{undef.})$

8 Reference points: $(0, \text{undef.})$, $\left(\dfrac{3\pi}{4}, 2\right)$, $\left(\dfrac{3\pi}{2}, \text{undef.}\right)$, $\left(\dfrac{9\pi}{4}, -2\right)$, $(3\pi, \text{undef.})$

9 Reference points: $\left(-\dfrac{1}{\pi}, \text{undef.}\right)$, $\left(\dfrac{1}{2} - \dfrac{1}{\pi}, 1\right)$, $\left(1 - \dfrac{1}{\pi}, \text{undef.}\right)$, $\left(\dfrac{3}{2} - \dfrac{1}{\pi}, -1\right)$, $\left(2 - \dfrac{1}{\pi}, \text{undef.}\right)$

10 Reference points: $\left(-\dfrac{\pi}{3}, \text{undef.}\right)$, $\left(-\dfrac{\pi}{6}, 1\right)$, $(0, \text{undef.})$, $\left(\dfrac{\pi}{6}, -1\right)$, $\left(\dfrac{\pi}{3}, \text{undef.}\right)$

11 Reference points: $(-\pi, \text{undef.})$, $\left(-\frac{\pi}{2}, \sqrt{3} - 2\right)$, $(0, \text{undef.})$,

$\left(\frac{\pi}{2}, \sqrt{3} - \sqrt{2}\right)$, $(\pi, \text{undef.})$

12 Reference points: $\left(-\frac{\pi}{2}, \text{undef.}\right)$, $\left(\frac{\pi}{4}, 0\right)$, $(\pi, \text{undef.})$, $\left(\frac{7\pi}{4}, 4\right)$,

$\left(\frac{5\pi}{2}, \text{undef.}\right)$

13 Reference points: $(2\pi, \text{undef.})$, $\left(\frac{5\pi}{2}, 2\right)$, $(3\pi, \text{undef.})$, $\left(\frac{7\pi}{2}, -4\right)$,

$(4\pi, \text{undef.})$

14 Reference points: $\left(-\frac{3}{2}, \text{undef.}\right)$, $\left(\frac{\pi}{4} - \frac{3}{2}, \frac{1}{2}\right)$, $\left(\frac{\pi}{2} - \frac{3}{2}, \text{undef.}\right)$,

$\left(\frac{3\pi}{4} - \frac{3}{2}, \frac{3}{2}\right)$, $\left(\frac{5\pi}{2} - \frac{3}{2}, \text{undef.}\right)$

15 Reference points: $(-\pi, \text{undef.})$, $\left(-\frac{\pi}{2}, 1\right)$, $(0, \text{undef.})$, $\left(\frac{\pi}{2}, 5\right)$,

$(\pi, \text{undef.})$

Review Exercise

1 Reference points: $(0, 0)$, $(2\pi, \sqrt{3})$, $(4\pi, 0)$, $(6\pi, -\sqrt{3})$, $(8\pi, 0)$

2 Reference points: $\left(-\frac{\pi}{6}, \text{undef.}\right)$, $\left(-\frac{\pi}{24}, -1\right)$, $\left(\frac{\pi}{12}, 0\right)$, $\left(\frac{5\pi}{24}, 1\right)$,

$\left(\frac{\pi}{3}, \text{undef.}\right)$

3 Reference points: $(0, \text{undef.})$, $\left(\frac{\pi}{14}, -1\right)$, $\left(\frac{\pi}{7}, \text{undef.}\right)$, $\left(\frac{3\pi}{7}, 1\right)$,

$\left(\frac{2\pi}{7}, \text{undef.}\right)$

4 Reference points: $(0, \text{undef.})$, $\left(\frac{1}{8}, 5\right)$, $\left(\frac{1}{4}, 0\right)$, $\left(\frac{3}{8}, -5\right)$, $\left(\frac{1}{2}, \text{undef.}\right)$

5 Reference points: $\left(\frac{1}{3}, 1\right)$, $\left(\frac{\pi}{6} + \frac{1}{3}, 0\right)$, $\left(\frac{\pi}{3} + \frac{1}{3}, -1\right)$, $\left(\frac{\pi}{2} + \frac{1}{3}, 0\right)$,

$\left(\frac{2\pi}{3} + \frac{1}{3}, 1\right)$

6 Reference points: $(0, k)$, $\left(\frac{9\pi}{2}, \text{undef.}\right)$, $(9\pi, -k)$, $\left(\frac{27\pi}{2}, \text{undef.}\right)$,

$\left(-\frac{9\pi}{2}, \text{undef.}\right)$

7 Reference points: $\left(-\frac{\pi}{8}, \text{undef.}\right)$, $\left(\frac{\pi}{8}, \frac{2}{3}\right)$, $\left(\frac{3\pi}{8}, 0\right)$, $\left(\frac{5\pi}{8}, -\frac{2}{3}\right)$, $\left(\frac{7\pi}{8}, \text{undef.}\right)$

8 Reference points: $\left(-2 - \frac{\pi}{2}, \text{undef.}\right)$, $\left(-2 - \frac{\pi}{4}, -1\right)$, $(-2, 0)$, $\left(\frac{\pi}{4} - 2, 1\right)$, $\left(\frac{\pi}{2} - 2, \text{undef.}\right)$

9 Reference points: $(0, -1)$, $\left(\frac{\pi}{10}, 0\right)$, $\left(\frac{\pi}{5}, 1\right)$, $\left(\frac{3\pi}{10}, 0\right)$, $\left(\frac{2\pi}{5}, -1\right)$

10 Reference points: $(5\pi, 0)$, $\left(\frac{15\pi}{2}, \frac{1}{7}\right)$, $(10\pi, 0)$, $\left(\frac{25\pi}{2}, -\frac{1}{7}\right)$, $(15\pi, 0)$

11 Reference points: $(0, k)$, $\left(\frac{3\pi}{2}, 0\right)$, $(3\pi, -k)$, $\left(\frac{9\pi}{2}, 0\right)$, $(6\pi, k)$

12 Reference points: $(0, 0)$, $\left(\frac{3\pi}{2}, k\right)$, $(3\pi, 0)$, $\left(\frac{9\pi}{2}, -k\right)$, $(6\pi, 0)$

Inverse Trigonometric Functions

7

7/Inverse Trigonometric Functions

7-1 Introduction Before defining the inverse trigonometric functions, let us review functions, relations, and their inverses with the following simple example.

The set of ordered pairs $h = \{(1,3),(2,3),(3,4)\}$ is a mathematical *relation* with domain $D_h = \{1,2,3\}$ and range $R_h = \{3,4\}$. Since h contains no two different ordered pairs having the same first coordinate, h is a *function*. (For a review of functions and relations, consult Appendix A.) Now we can form the *inverse relation* to h, $k = \{(3,1),(3,2),(4,3)\}$, by interchanging the coordinates of h. The domain of k is $D_k = \{3,4\}$, and the range of k is $R_k = \{1,2,3\}$. Note that k is a mathematical relation, but *not* a function since the ordered pairs $(3,1)$ and $(3,2)$ have the same first element.

Since we wish to concentrate our efforts on the study of functions, we will have to restrict the domain of h in order to ensure that the inverse relation we form will also be a function. We can choose a subset H of h such that $H = \{(2,3),(3,4)\}$. We call this the *principal value function* of h. Now if $K = \{(3,2),(4,3)\}$, K is the *inverse* relation for H, but K is also a function. In the next section we will use the same technique to define the inverse trigonometric functions.

Exercise 7-1.1 For the following state whether each is or is not a function:

1 $G = \{(1,2),(1,3),(2,3),(3,4)\}$

2 $H = \{(2,1),(3,1),(4,1)\}$

3 $J = \{(x,y)|y^2 = x^2\}$

4 $K = \{(x,y)|y = 3x^2\}$

Choose a subset F of each of the following such that F^{-1} is a function:

5 $f = \{(2,2),(3,3),(4,4),(4,5)\}$ **6** $g = \{(1,3),(1,4),(1,5)\}$

7 $h = \{(x,y)|y = x^2\}$ **8** $k = \{(x,y)|x^2 + y^2 = 4\}$

7-2 Inverse Sine Function

The set $f = \{(x,y)|y = \sin x\}$ defines a mathematical relation with domain R (all real numbers) and range $-1 \le x \le 1$. And f is also a function since no two different ordered pairs have the same first coordinate. If we form $g = \{(y,x)|y = \sin x\}$, g is the inverse relation for f, but g is *not* a function. Using the technique described in Section 7-1, we can restrict the domain of f and consider the subset $F = \left\{(x,y)|y = \sin x, -\dfrac{\pi}{2} \le x \le \dfrac{\pi}{2}\right\}$. Then F is called the "principal value function" of $y = \sin x$. Now if we form $F^{-1} = \left\{(y,x)|y = \sin x, -\dfrac{\pi}{2} \le x \le \dfrac{\pi}{2}\right\}$, then F^{-1} will be the inverse *function* for F. When writing F and F^{-1} without using set notation, we write $y = \text{Sin } x$ for F (called the "cap-sine function" for obvious reasons) and $y = \arcsin x$, or $y = \text{Sin}^{-1} x$ for F^{-1}. Note that $y = \arcsin x$ and $y = \text{Sin}^{-1} x$ name the same function and have the following meaning: y is the angle (number) whose sine is x. Table 7-2.1 summarizes the arcsine function. Figure 7-2.1 shows the graph of $y = \arcsin x$.

Table 7-2.1

Function	Domain	Range
$y = \text{Sin}^{-1} x$ $y = \arcsin x$ $(\sin y = x)$	$\{x \mid -1 \le x \le 1\}$	$\left\{y \mid -\dfrac{\pi}{2} \le y \le \dfrac{\pi}{2}\right\}$

Figure 7-2.1

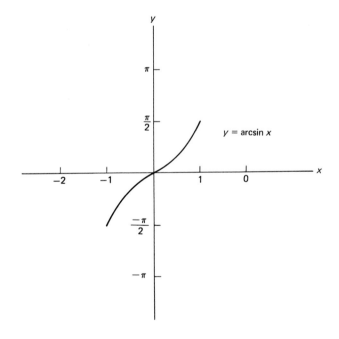

Exercise 7-2.1 Give the domain and range of the following:

1 $y = \sin x$

2 $y = \text{Sin } x$

3 $y = \text{Sin}^{-1}(x)$

7-3 Inverses of Other Trigonometric Functions By using our technique of restricting the domain of the function which we wish to consider, we can define the principal value function for each of the other trigonometric functions in the following manner:

$$y = \begin{cases} \text{Cos } x & \text{if } y = \cos x \text{ and } 0 \le x \le \pi \\[2mm] \text{Tan } x & \text{if } y = \tan x \text{ and } -\dfrac{\pi}{2} < x < \dfrac{\pi}{2} \\[2mm] \text{Cot } x & \text{if } y = \cot x \text{ and } 0 < x < \pi \\[2mm] \text{Sec } x & \text{if } y = \sec x \text{ and } 0 \le x \le \pi,\, x \ne \dfrac{\pi}{2} \\[2mm] \text{Csc } x & \text{if } y = \csc x \text{ and } -\dfrac{\pi}{2} \le x \le \dfrac{\pi}{2},\, x \ne 0 \end{cases}$$

Now by switching the coordinates we form the inverse trigonometric functions $\text{Cos}^{-1} x$, $\text{Tan}^{-1} x$, $\text{Cot}^{-1} x$, $\text{Sec}^{-1} x$, and $\text{Csc}^{-1} x$. Table 7-3.1 summarizes the inverse trigonometric functions.

Table 7-3.1

The inverse trigonometric functions

Function	Domain	Range
$y = \text{Arcsin } x$ $y = \text{Sin}^{-1} x$ $(\sin y = x)$	$-1 \le x \le 1$	$-\dfrac{\pi}{2} \le y \le \dfrac{\pi}{2}$
$y = \text{Arccos } x$ $y = \text{Cos}^{-1} x$ $(\cos y = x)$	$-1 \le x \le 1$	$0 \le y \le \pi$
$y = \text{Arctan } x$ $y = \text{Tan}^{-1} x$ $(\tan y = x)$	R (all real numbers)	$-\dfrac{\pi}{2} < y < \dfrac{\pi}{2}$
$y = \text{Arccot } x$ $y = \text{Cot}^{-1} x$ $(\cot y = x)$	R	$0 < y < \pi$

Table 7-3.1
(Continued)

Function	Domain	Range
$y = $ Arcsec x $y = $ Sec^{-1} x (sec $y = x$)	$x \leq -1$ or $1 \leq x$	$0 \leq y \leq \pi$ $y \neq \dfrac{\pi}{2}$
$y = $ Arccsc x $y = $ Csc^{-1} x (csc $y = x$)	$x \leq -1$ or $1 \leq x$	$-\dfrac{\pi}{2} \leq y \leq \dfrac{\pi}{2}$ $y \neq 0$

The graph of $y = $ Sin^{-1} x was given in Figure 7-2.1. Figures 7-3.1 through 7-3.5 show the graphs of the remaining inverse trigonometric functions.

Figure 7-3.1

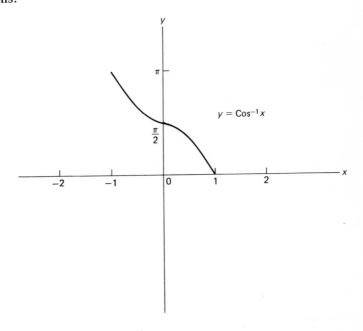

Figure 7-3.2

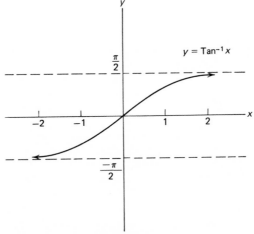

Figure 7-3.3

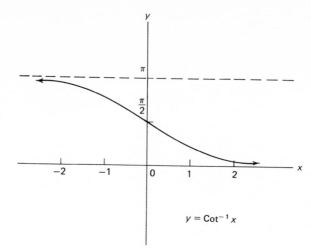

$y = \text{Cot}^{-1} x$

Figure 7-3.4

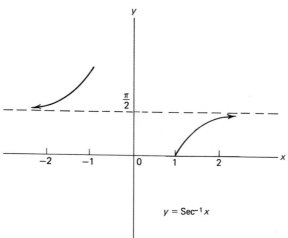

$y = \text{Sec}^{-1} x$

Figure 7-3.5

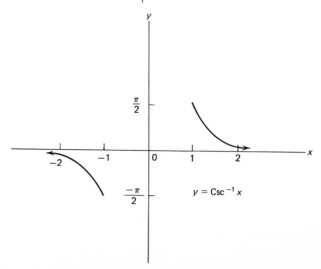

$y = \text{Csc}^{-1} x$

Exercise 7-3.1 Define the principal value function of:

1 $y = \cos x$

2 $y = \tan x$

3 $y = \cot x$

4 $y = \sec x$

5 $y = \csc x$

Give the domain and range for:

6 $y = \cos x$

7 $y = \operatorname{Cot}^{-1} x$

8 $y = \operatorname{Tan}^{-1} x$

9 $y = \operatorname{Sec}^{-1} x$

10 $y = \csc x$

11 $y = \sec x$

12 $y = \cot x$

13 $y = \operatorname{Csc}^{-1} x$

7-4 Problems Involving Inverse Trigonometric Functions

In addition to the sketching of graphs of the inverse trigonometric functions, it is often necessary to solve problems of the following kind:

Given $y = T^{-1}(x)$ (where T is an abbreviation for a principle value of a trigonometric function) and a value for x, find y.

Given $y = t[T^{-1}(x)]$ (where t is an abbreviation for a trigonometric function) and a value for x, find y.

The following examples show a method for solving problems of the type $y = T^{-1}(x)$.

Example 7-4.1 Given $y = \text{Sin}^{-1} \dfrac{\sqrt{3}}{2}$, find y.

Solution Since y is the angle between $-\dfrac{\pi}{2}$ and $\dfrac{\pi}{2}$ such that $\sin y = \dfrac{\sqrt{3}}{2}$, from Section 2-2 we get $y = \dfrac{\pi}{3}$.

Example 7-4.2 Given $y = \text{Tan}^{-1}\left(-\dfrac{1}{\sqrt{3}}\right)$, find y

Solution Since y is the angle between $-\dfrac{\pi}{2}$ and $\dfrac{\pi}{2}$ such that $\tan y = -\dfrac{1}{\sqrt{3}}$, from Section 2-2 we see that $y = -\dfrac{\pi}{6}$.

Example 7-4.3 Given $y = \text{Sec}^{-1} 3.079$, find y.

Solution Since y is the angle such that $\sec y = 3.079$, from Table C-2 we see $y = 1.24$.

Example 7-4.4 Given $y = \text{Csc}^{-1} 1.444$, find y.

Solution Since y is the angle such that $\csc y = 1.444$, from Table C-1 we see $y = 43°50'$.

To find y when given $y = T^{-1}(x)$ and a value for $x = a$, we look in the appropriate table for the value of the angle (y) having $T(y) = a$, always making sure our y value lies in the domain of T.

The next four examples show a method for finding y, given $y = t[T^{-1}(x)]$.

Example 7-4.5   Given $y = \sin [\arctan (-1)]$, find y.

Solution   Let $s = \arctan (-1)$. Then, $\tan s = -1$ and $-\dfrac{\pi}{2} < s < \dfrac{\pi}{2}$. Now, from Section 2-2, $\tan \left(-\dfrac{\pi}{4}\right) = -1$ and $s = -\dfrac{\pi}{4}$. Since $\sin \left(-\dfrac{\pi}{4}\right) = -\dfrac{\sqrt{2}}{2}$, we have $y = -\dfrac{\sqrt{2}}{2}$.

Example 7-4.6   Given $y = \sec (\text{Tan}^{-1} 2)$, find y.

Solution   Let $s = \text{Tan}^{-1} 2$. Construct a reference triangle. Now, $\tan s = \dfrac{a}{b} = \dfrac{2}{1}$; hence $c = \sqrt{5}$. Since $y = \sec s = \dfrac{c}{b} = \dfrac{\sqrt{5}}{1}$, it follows that $y = \sqrt{5}$. See Figure 7-4.1.

Figure 7-4.1

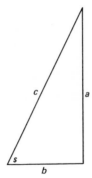

Example 7-4.7   Given $y = \tan \left(\text{arcsec} \dfrac{\sqrt{1 + x^2}}{x}\right)$, $x \neq 0$, find y.

Solution   Let s be the angle such that $\sec s = \dfrac{\sqrt{1 + x^2}}{x}$, $x \neq 0$. Construct a reference triangle. Then $y = \tan s = \dfrac{1}{x}$. See Figure 7-4.2.

Figure 7-4.2

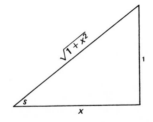

Example 7-4.8 Given $y = \sin\left(\mathrm{Sec}^{-1}\dfrac{x}{2}\right)$, $x > 2$, find y.

Solution Let s be the angle such that $\sec s = \dfrac{x}{2}$, $x > 2$. Construct a reference triangle. Then $y = \sin s = \dfrac{\sqrt{x^2 - 4}}{x}$, $x > 2$. See Figure 7-4.3.

Figure 7-4.3

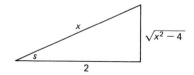

The above illustrations show the following steps:

STEP 1 Name the angle whose functional value is given.

STEP 2 (a) Using the given functional value, find the measure of the angle. Be careful to choose the angle that belongs to the appropriate principal value function.
(b) Construct a reference triangle.
(c) Give the required functional value from the above step.

Exercise 7-4.1 **1** Complete the following table. Write "Undef." for any functional value that is undefined. Give approximations from a table if necessary.

x	$\mathrm{Sin}^{-1} x$	$\mathrm{Cos}^{-1} x$	$\mathrm{Tan}^{-1} x$	$\mathrm{Cot}^{-1} x$	$\mathrm{Sec}^{-1} x$	$\mathrm{Csc}^{-1} x$
0						
1						
$-\dfrac{1}{2}$						
$\dfrac{\sqrt{3}}{2}$						
$-\dfrac{\sqrt{2}}{2}$						

Find *y* for each of the following:

2 $y = \text{Sin}^{-1}\left(-\dfrac{\sqrt{3}}{2}\right)$

3 $y = \text{Tan}^{-1}(-1)$

4 $y = \text{Cot}^{-1}\dfrac{1}{\sqrt{3}}$

5 $y = \text{Sec}^{-1}(-2)$

6 $y = \text{arccsc } 2$

7 $y = \text{Cos}^{-1}\dfrac{\sqrt{2}}{2}$

8 $y = \arcsin\left(-\dfrac{\sqrt{2}}{2}\right)$

9 $y = \text{Sec}^{-1}\dfrac{2}{\sqrt{3}}$

10 $y = \text{Sin}^{-1} 0$

11 $y = \text{Csc}^{-1} 0$

Find *y* for each of the following:

12 $y = \sin\left(\text{Tan}^{-1}\dfrac{x}{2}\right),\ x > 0$

13 $y = \tan\left(\text{Sin}^{-1}\dfrac{x}{\sqrt{x^2 + 9}}\right),\ x > 0$

14 $y = \sin\left(\text{Sin}^{-1}\tfrac{3}{5}\right)$

15 $y = \tan\left(\text{Sin}^{-1}\tfrac{3}{5}\right)$

16 $y = \csc\left(\text{Sin}^{-1}\tfrac{3}{5}\right)$

17 $y = \tan\left[\text{Cos}^{-1}\left(-\tfrac{15}{17}\right)\right]$

18 $y = \sin\left(\text{Sec}^{-1}\tfrac{29}{21}\right)$

19 $y = \sec [\text{arccsc} (-\frac{17}{8})]$

20 $y = \cos [\text{arccsc} (-\frac{29}{21})]$

Review Exercise **1** State whether each of the following is or is not a function.

(a) $A = \{(3,2),(7,1),(4,4),(5,4)\}$

(b) $S = \{(\pi,3),(6,3),(7,3),(5,3)\}$

(c) $H = \{(2,\frac{1}{2}),(3,\frac{1}{3}),\left(\pi,\frac{1}{\pi}\right),(0,1)\}$

(d) $K = \{(2,3),(2,4),(2,5),(2,7)\}$

(e) $W = \{(x,y)|y = 3x + 2\}$

(f) $T = \{(x,y)|y^2 = x\}$

2 Write the inverse relation for each set in problem 1.

3 Identify those relations in problem 2 that are also functions.

4 Choose a subset F of each of the following mathematical relations such that F^{-1} is a function.

(a) $f = \{(1,1),(2,1),(1,5),(2,6)\}$

(b) $g = \{(a,b),(a,c),(b,c),(c,d)\}$

(c)　$h = \{(\pi,5),(2\pi,10),(\pi,7),(3\pi,4)\}$

(d)　$i = \{(x,y)|y = \pm x\}$

5　Give the domain and range of:

(a)　$y = \sin x$

(b)　$y = \text{Sin}^{-1} x$

6　Sketch the graph of:

(a)　$y = 2 \sin 2x$

(b)　$y = 2 \text{Sin}^{-1} 2x$

(c)　$y = \text{Arctan } x$

(d)　$y = \text{Arcsec } x$

7　Solve for y.

(a)　$y = \text{Cos}^{-1} \dfrac{\sqrt{3}}{2}$
　　　　　　　　　　　　　　　(b)　$y = \text{Sin}^{-1} \dfrac{\sqrt{2}}{2}$

(c)　$y = \text{Tan}^{-1} 1$
　　　　　　　　　　　　　　　(d)　$y = \text{Cos}^{-1} \left(-\dfrac{\sqrt{2}}{2}\right)$

(e)　$y = \text{Sec}^{-1} 2$
　　　　　　　　　　　　　　　(f)　$y = \text{Csc}^{-1} \frac{1}{2}$

(g)　$y = \text{arccsc } \dfrac{\sqrt{3}}{2}$

8 Solve for y.

(a) $y = \sin (\text{Sin}^{-1} 1)$

(b) $y = \cos (\text{Sin}^{-1} \sqrt{2})$

(c) $y = \tan (\text{Sin}^{-1} \sqrt{3})$

(d) $y = \csc \left(\text{Cos}^{-1} \dfrac{1}{\sqrt{2}} \right)$

(e) $y = \sec (\text{Tan}^{-1} 1)$

(f) $y = \sec (\text{Cot}^{-1} 1)$

(g) $y = \cos \left(2 \arcsin \dfrac{1}{\sqrt{2}} \right)$

(h) $y = \cot (\text{arcsec } \sqrt{2})$

(i) $y = \cos (\arcsin \tfrac{3}{5})$

(j) $y = \csc [\arccos (-\tfrac{15}{17})]$

Answers

Exercise 7-1.1

1 Not a function

2 A function

3 Not a function

4 A function

5 $F = \{(2,2),(3,3)\}$

6 $F = \{(1,3)\}$

7 $F = \{(x,y)|y = x^2,\ x \geq 0\}$

8 $F = \{(x,y)|y = \sqrt{4 - x^2},\ -2 \leq x \leq 2\}$

Exercise 7-2.1

1 Domain $\{x|x \in R\}$, Range $\{y|-1 \leq y \leq 1\}$

2 Domain $\left\{x \left| -\dfrac{\pi}{2} \leq x \leq \dfrac{\pi}{2} \right. \right\}$, Range $\{y|-1 \leq y \leq 1\}$

3 Domain $\{x|-1 \leq x \leq 1\}$, Range $\left\{y \left| -\dfrac{\pi}{2} \leq y \leq \dfrac{\pi}{2} \right. \right\}$

Exercise 7-3.1

1 $y = \text{Cos } x$, Domain $\{x|0 \leq x \leq \pi\}$

2 $y = \text{Tan } x$, Domain $\left\{x \left| -\dfrac{\pi}{2} < x < \dfrac{\pi}{2} \right. \right\}$

3 $y = \text{Cot } x$, Domain $\{x|0 < x < \pi\}$

4 $y = \text{Sec } x$, Domain $\left\{x \mid 0 \le x \le \pi, \ x \ne \dfrac{\pi}{2}\right\}$

5 $y = \text{Csc } x$, Domain $\left\{x \mid -\dfrac{\pi}{2} \le x \le \dfrac{\pi}{2}, \ x \ne 0\right\}$

6 Domain $\{x \mid x \in R\}$, Range $\{y \mid -1 \le y \le 1\}$

7 Domain $\{x \mid x \in R\}$, Range $\{y \mid 0 < y < \pi\}$

8 Domain $\{x \mid x \in R\}$, Range $\left\{y \mid -\dfrac{\pi}{2} < y < \dfrac{\pi}{2}\right\}$

9 Domain $\{x \mid x \ge 1, \ x \le -1\}$, Range $\left\{y \mid 0 \le y \le \pi, \ y \ne \dfrac{\pi}{2}\right\}$

10 Domain $\{x \mid x \in R, \ x \ne n\pi, \ n = 0, 1, 2, 3, \ldots\}$
Range $\{y \mid y \ge 1, \ y \le -1\}$

11 Domain $\left\{x \mid x \in R, \ x \ne \dfrac{(2n + 1)\pi}{2}, \ n = 0, 1, 2, 3, \ldots\right\}$
Range $\{y \mid y \ge 1, \ y \le -1\}$

12 Domain $\{x \mid x \in R, \ x \ne n\pi, \ n = 0, 1, 2, 3, \ldots\}$
Range $\{y \mid y \in R\}$

13 Domain $\{x \mid x \ge 1, \ x \le -1\}$
Range $\left\{y \mid -\dfrac{\pi}{2} \le y \le \dfrac{\pi}{2}, \ y \ne 0\right\}$

Exercise 7-4.1 **1**

x	$\text{Sin}^{-1} x$	$\text{Cos}^{-1} x$	$\text{Tan}^{-1} x$	$\text{Cot}^{-1} x$	$\text{Sec}^{-1} x$	$\text{Csc}^{-1} x$
0	0	$\dfrac{\pi}{2}$	0	$\dfrac{\pi}{2}$	Undef.	Undef.
1	$\dfrac{\pi}{2}$	0	$\dfrac{\pi}{4}$	$\dfrac{\pi}{4}$	0	$\dfrac{\pi}{2}$
$-\dfrac{1}{2}$	$-\dfrac{\pi}{6}$	$\dfrac{2\pi}{3}$	-0.46	2.03	Undef.	Undef.
$\dfrac{\sqrt{3}}{2}$	$\dfrac{\pi}{3}$	$\dfrac{\pi}{6}$	0.72	0.86	Undef.	Undef.
$-\dfrac{\sqrt{2}}{2}$	$-\dfrac{\pi}{4}$	$\dfrac{3\pi}{4}$	-0.62	2.18	Undef.	Undef.

2 $-\dfrac{\pi}{3}$ **3** $-\dfrac{\pi}{4}$ **4** $\dfrac{\pi}{3}$ **5** $\dfrac{2\pi}{3}$

6 $\dfrac{\pi}{6}$ **7** $\dfrac{\pi}{4}$ **8** $-\dfrac{\pi}{4}$ **9** $\dfrac{5\pi}{6}$

10 0   **11** Undef.   **12** $\dfrac{x}{\sqrt{x^2 + 4}}$ **13** $\dfrac{x}{3}$

14 $\dfrac{3}{5}$ **15** $\dfrac{3}{4}$ **16** $\dfrac{5}{3}$ **17** $-\dfrac{8}{15}$

18 $\dfrac{20}{29}$ **19** $-\dfrac{17}{15}$ **20** $\dfrac{20}{29}$

Review Exercise

1 (*a*) A function   (*b*) A function
    (*c*) A function   (*d*) Not a function
    (*e*) A function   (*f*) Not a function

2 (*a*) {(2,3),(1,7),(4,4),(4,5)}   (*b*) {(3,π),(3,6),(3,7),(3,5)}

    (*c*) $\left\{(\tfrac{1}{2},2),(\tfrac{1}{3},3),\left(\dfrac{1}{\pi},\pi\right),(1,0)\right\}$ (*d*) {(3,2),(4,2),(5,2),(7,2)}

    (*e*) $\left\{(x,y)\Big| y = \dfrac{x - 2}{3}\right\}$ (*f*) $\{(x,y)| y = x^2\}$

3 (*c*), (*d*), (*e*), (*f*)

4 (*a*) {(1,5),(2,6)}   (*b*) {(*a*,*b*),(*b*,*c*)}
    (*c*) {(π,5),(2π,10)}   (*d*) {(x,y)| y = −x}

5 (*a*) Domain {$x| x \in R$}, Range {$y| -1 \le y \le 1$}

    (*b*) Domain {$x| -1 \le x \le 1$}, Range $\left\{y\Big| -\dfrac{\pi}{2} \le y \le \dfrac{\pi}{2}\right\}$

6

    (*a*)

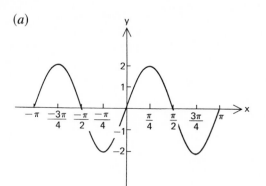

    (*b*)

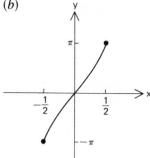

(c)

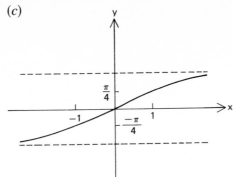

(d)

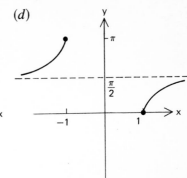

7 (a) $\dfrac{\pi}{6}$ (b) $\dfrac{\pi}{4}$ (c) $\dfrac{\pi}{4}$ (d) $\dfrac{3\pi}{4}$

   (e) $\dfrac{\pi}{3}$ (f) Undefined   (g) Undefined

8 (a) 1   (b) Undefined   (c) Undefined   (d) $\sqrt{2}$

   (e) $\sqrt{2}$ (f) $\sqrt{2}$ (g) 0   (h) 1

   (i) $\frac{4}{5}$ (j) $\frac{17}{8}$

Fundamental Trigonometric Identities

8

Objectives

Upon completion of Chapter 8, you will be able to:

1. **Write from memory the following identities: (8-2)**

   $\sin A \csc A = 1$

   $\cos A \sec A = 1$

   $\tan A \cot A = 1$

   $\sin^2 A + \cos^2 A = 1$

   $1 + \tan^2 A = \sec^2 A$

   $1 + \cot^2 A = \csc^2 A$

   $\tan A = \dfrac{\sin A}{\cos A}$

   $\cot A = \dfrac{\cos A}{\sin A}$

2. **Simplify expressions using the identities in objective 1. (8-3)**

3. **Prove various identities by using the identities in objective 1. (8-4)**

8/Fundamental Trigonometric Identities

8-1 Introduction An *identity* is an equation which is true for all permissible values of the variables. A *conditional equation* is an equation which is true for selected values of the variable but is false for other values of the variable.

Example 8-1.1 $256x^4y^4 - 289z^2 = (16x^2y^2 + 17z)(16x^2y^2 - 17z)$

This is an identity.

Example 8-1.2 $x^2 + 3x + 2 = 0$ is a conditional equation.

Example 8-1.3 $\sin \theta = \dfrac{1}{\csc \theta}$ is an identity. It will be proved later.

Example 8-1.4 $\cos \theta = \frac{1}{2}$ is a conditional equation, and methods of solving for θ will be shown later.

8-2 Fundamental Identities Some trigonometric identities are easily proved directly from the definitions. Three of these are called the *reciprocal* identities. They are

8-2.1 $\sin \theta \csc \theta = 1$

8-2.2 $\cos \theta \sec \theta = 1$

8-2.3 $\tan \theta \cot \theta = 1$

To prove $\tan \theta \cot \theta = 1$, we simply recall that $\tan \theta = \dfrac{y}{x}$ and $\cot \theta = \dfrac{x}{y}$ from the definitions in Chapter 3. The proof is written formally as follows:

Proof
$$\tan \theta \cot \theta = \frac{y}{x} \cdot \frac{x}{y}$$

$$= \frac{yx}{xy} = \frac{xy}{xy}$$

$$= 1$$

Identities 8-2.1 and 8-2.2 are proved similarly.

The identities

8-2.4 $\tan \theta = \dfrac{\sin \theta}{\cos \theta}$

8-2.5 $\cot \theta = \dfrac{\cos \theta}{\sin \theta}$

are called the *ratio* identities and can also be proved directly from the definitions.

The squared identities

8-2.6 $\sin^2 \theta + \cos^2 \theta = 1$

8-2.7 $1 + \tan^2 \theta = \sec^2 \theta$

8-2.8 $1 + \cot^2 \theta = \csc^2 \theta$

are proved by considering θ to be any angle in standard position, (x,y) to be on the terminal side of θ, and r to be the length from $(0,0)$ to (x,y). Hence, identity 8-2.7 is derived as follows:

$$1 + \tan^2 \theta = 1 + \left(\frac{y}{x}\right)^2$$

$$= 1 + \frac{y^2}{x^2}$$

$$= \frac{x^2 + y^2}{x^2}$$

$$= \frac{r^2}{x^2} = \left(\frac{r}{x}\right)^2$$

$$= \sec^2 \theta$$

Identities 8-2.6 and 8-2.8 are derived similarly.

8-3 Simplifying Trigonometric Expressions

Identities 8-2.1 through 8-2.8 are extremely important and should be memorized. They will be used extensively to simplify trigonometric expressions and to prove other identities. The following examples illustrate how trigonometric expressions may be simplified.

Example 8-3.1   Simplify

$$\frac{\sin x \csc x}{\tan x} + \frac{\cos x}{\sin x}$$

Solution
$$\frac{\sin x \csc x}{\tan x} + \frac{\cos x}{\sin x} = \frac{1}{\tan x} + \cot x$$

$$= \cot x + \cot x$$

$$= 2 \cot x$$

Example 8-3.2   Simplify

$$\frac{\cos A}{\sin A} + \frac{\sin A}{\cos A} + \frac{\sec A}{\sin A}$$

Solution
$$\frac{\cos A}{\sin A} + \frac{\sin A}{\cos A} + \frac{\sec A}{\sin A} = \frac{\cos^2 A + \sin^2 A + \sec A \cos A}{\sin A \cos A}$$

$$= \frac{(\cos^2 A + \sin^2 A) + (\sec A \cos A)}{\sin A \cos A}$$

$$= \frac{1 + 1}{\sin A \cos A}$$

$$= \frac{2}{\sin A \cos A}$$

$$= 2 \cdot \frac{1}{\sin A} \cdot \frac{1}{\cos A}$$

$$= 2 \csc A \sec A$$

Example 8-3.3   Simplify

$$\frac{1}{\tan x + \cot x}$$

Solution
$$\frac{1}{\tan x + \cot x} = \frac{1}{\dfrac{\sin x}{\cos x} + \dfrac{\cos x}{\sin x}}$$

$$= \frac{1}{\dfrac{\sin^2 x + \cos^2 x}{\cos x \sin x}}$$

$$= \frac{1}{\dfrac{1}{\cos x \sin x}}$$

$$= \cos x \sin x$$

Example 8-3.4   Simplify $1 - 2 \cos^2 x + \cos^4 x$.

Solution $\begin{aligned} 1 - 2 \cos^2 x + \cos^4 x &= (1 - \cos^2 x)^2 \\ &= (\sin^2 x)^2 \\ &= \sin^4 x \end{aligned}$

Exercise 8-3.1   **1**   Simplify each of the following:

(a) $\csc t - \csc t \cos^2 t$

(b) $\tan^2 A \cos^2 A + \cot^2 A \sin^2 A$

(c) $(1 - \sin^2 s)\sec^2 s$

(d) $\tan x \sec^2 x - \tan^3 x$

(e) $\dfrac{\sec^2 x - 1}{\csc^2 x - 1}$

(f) $(\cos A + \sin A)^2 + (\sin A - \cos A)^2$

2   Solve $\cos^2 x + \sin^2 x = 1$ for $\sin x$.

3   Solve $\cos^2 x + \sin^2 x = 1$ for $\cos x$.

4   Solve $1 + \tan^2 A = \sec^2 A$ for $\tan A$.

5   Test the identity $\tan x = \dfrac{\sin x}{\cos x}$ for $x = 30°$ and $x = 60°$.

8-4 Proving Identities One method for proving or verifying identities makes use of previously known identities. Through substitutions and algebraic processes one member is transformed into the other member, hence proving the identity. The following examples illustrate this method.

Example 8-4.1 Prove the identity

$$\frac{\sin t}{1 - \cos t} - \cot t = \csc t \qquad \text{if } \cos t \neq 1$$

Proof
$$\frac{\sin t}{1 - \cos t} - \cot t = \left(\frac{\sin t}{1 - \cos t} \cdot \frac{1 + \cos t}{1 + \cos t}\right) - \cot t$$

$$= \frac{\sin t(1 + \cos t)}{1 - \cos^2 t} - \frac{\cos t}{\sin t}$$

$$= \frac{\sin t(1 + \cos t)}{\sin^2 t} - \frac{\cos t}{\sin t}$$

$$= \frac{1 + \cos t}{\sin t} - \frac{\cos t}{\sin t}$$

$$= \frac{1 + \cos t - \cos t}{\sin t}$$

$$= \frac{1}{\sin t}$$

$$= \csc t$$

Example 8-4.2 Prove the identity

$$\tan^3 x - \cot^3 x = (\tan x - \cot x)(\sec^2 x + \cot^2 x)$$

Proof
$$\tan^3 x - \cot^3 x = (\tan x - \cot x)(\tan^2 x + \tan x \cot x + \cot^2 x)$$

$$= (\tan x - \cot x)(\tan^2 x + 1 + \cot^2 x)$$

$$= (\tan x - \cot x)[(\tan^2 x + 1) + \cot^2 x]$$

$$= (\tan x - \cot x)(\sec^2 x + \cot^2 x)$$

Some possible helpful hints on proving identities are:

1 Work with the more complicated side by making correct substitutions.
2 Know what expression you are trying to obtain on the other side; i.e., you may wish to write out several different forms of that side on scratch paper.

3   It often helps to change the expression on the side with which you are working to sines and cosines.

4   Whenever fractions are involved, you may wish to try multiplication of numerator and denominator by the same expression, noting restrictions which make the variable zero.

Notice how some of these hints are used in the following examples.

Example 8-4.3   Prove that

$$\frac{1}{\tan A + \cot A} = \frac{1}{\sec A \csc A}$$

is an identity.

Proof $$\frac{1}{\tan A + \cot A} = \frac{1}{\dfrac{\sin A}{\cos A} + \dfrac{\cos A}{\sin A}}$$

$$= \frac{1}{\dfrac{\sin^2 A + \cos^2 A}{\cos A \sin A}}$$

$$= \frac{\cos A \sin A}{\sin^2 A + \cos^2 A}$$

$$= \frac{\cos A \sin A}{1}$$

$$= \frac{1}{\sec A} \cdot \frac{1}{\csc A}$$

$$= \frac{1}{\sec A \csc A}$$

Example 8-4.4   Prove the identity

$$(\csc x + \sec x)^2 = \sec^2 x \csc^2 x + 2 \sec x \csc x$$

Proof $\sec^2 x \csc^2 x + 2 \sec x \csc x = \left(\dfrac{1}{\cos x}\right)^2\left(\dfrac{1}{\sin x}\right)^2 + 2\left(\dfrac{1}{\sin x}\right)\left(\dfrac{1}{\cos x}\right)$

$$= \frac{1}{\cos^2 x \sin^2 x} + \frac{2}{\sin x \cos x}$$

$$= \frac{1 + 2 \sin x \cos x}{\sin^2 x \cos^2 x}$$

$$= \frac{(\cos^2 x + \sin^2 x) + 2 \sin x \cos x}{\sin^2 x \cos^2 x}$$

$$= \frac{\cos^2 x}{\sin^2 x \cos^2 x} + \frac{\sin^2 x}{\sin^2 x \cos^2 x} + \frac{2 \sin x \cos x}{\sin^2 x \cos^2 x}$$

$$= \frac{1}{\sin^2 x} + \frac{1}{\cos^2 x} + \frac{2}{\sin x \cos x}$$

$$= \csc^2 x + \sec^2 x + 2 \sec x \csc x$$

$$= (\csc x + \sec x)^2$$

Exercise 8-4.1   Verify the following identities:

1 $\dfrac{\sin x}{\csc x} + \dfrac{\cos x}{\sec x} = 1$

2 $\sec^2 A(1 - \sin^2 A) = 1$

3 $\dfrac{\cos B}{\sin B} + \dfrac{1 - \cos B}{\sin B} = \csc B$

4 $\dfrac{\sin C}{1 + \cos C} + \dfrac{1 + \cos C}{\sin C} = 2 \csc C$

5 $\dfrac{\cos y(1 + \cos y) + \sin^2 y}{\sin y(1 + \cos y)} = \csc y$

6 $(\tan A + \cot A)(\tan^2 A - \tan A \cot A + \cot^2 A) = \tan^3 A + \cot^3 A$

7 $(\csc x - \sec x)(\csc x + \sec x) = \dfrac{\cos^2 x - \sin^2 x}{\sin^2 x \cos^2 x}$

8 $\dfrac{1 + \tan^2 x}{\csc^2 x} = \tan^2 x$

9 $\dfrac{\sin B + 2 \sin B \cos B}{2 + \cos B - 2 \sin^2 B} = \tan B$

10 $\sec y + 2 \tan y = \dfrac{\cot y + 2 \cos y}{\csc y - \sin y}$

11 $\dfrac{1 + \cot^2 z}{\sec^2 z} = \cot^2 z$

12 $1 - \sin^4 x + \cos^4 x = 2 \cos^2 x$

13 $\dfrac{\sin A}{\csc A (1 + \cot^2 A)} = \sin^4 A$

14 $(\sin^3 x + \cos^3 x) = (\sin x + \cos x)(1 - \sin x \cos x)$

15 $\tan^2 x - \sin^2 x = \tan^2 x \sin^2 x$

16 $\dfrac{1}{\cos x \sin x} = \tan x + \cot x$

17 $\dfrac{1 - \sec y}{1 + \sec y} = \dfrac{\cos y - 1}{\cos y + 1}$

18 $\dfrac{1}{1 + \sin A} + \dfrac{1}{1 - \sin A} = 2 \sec^2 A$

19 $(\cos t - \sin t)(\sec t + \csc t) = \cot t - \tan t$

20 $\dfrac{\sin x}{\csc x} + \dfrac{\cos x}{\sec x} = \csc x \sin x$

21 $\dfrac{\sin^4 x - \cos^4 x}{\sin^2 x - \cos^2 x} = 1$

22 $\dfrac{1 + \sec^3 m}{1 + \sec m} = 1 - \sec m + \sec^2 m$

23 $\sin^2 x + \csc^2 x + \cos^2 x - \cot^2 x + \sec^2 x - \tan^2 x = 3$

Review Exercise **1** Fill in the blanks in each of the following:

(a) $\sin x \csc x =$ _____

(b) $\cos x \sec x =$ _____

(c) $\tan x \cot x =$ _____

(d) $\dfrac{\sin x}{\cos x} =$ _____

(e) $\dfrac{\cos x}{\sin x} =$ _____

(f) $1 + \tan^2 x =$ _____

(g) $1 + \cot^2 x =$ _____

(h) $\sin^2 x + \cos^2 x =$ _____

2 Simplify each of the following:

(a) $\dfrac{\sin^3 x}{\cos^3 x} + 5 \tan^3 x$

(b) $(1 - \sin x)(1 + \sin x)$

(c) $\sec^2 z + \tan^2 z$

(d) $\tan y \cot y - \cos^2 y$

(*e*) $\csc x \tan x - \sec x$

(*f*) $\dfrac{\sec^2 t - \tan^2 t}{\sin^2 t}$

3   Prove that each of the following is an identity:

(*a*) $\dfrac{\csc^2 A - \cot^2 A}{\sec^2 A} = \cos^2 A$

(*b*) $\dfrac{\cos^2 x \sin^2 x + \cos^4 x}{\sin^2 x} = \cot^2 x$

(*c*) $\cot B \cos B = \csc B - \sin B$

(*d*) $\dfrac{\cos^4 C - \sin^4 C}{\cos C + \sin C} = \cos C - \sin C$

(*e*) $\dfrac{\cos^2 x - \sin^2 x \cos^2 x}{\sin^2 x} = \cot^2 x \cos^2 x$

(*f*) $\tan^2 x(1 - \sin^2 x) + \cot^2 x = 1 + \dfrac{\cos^2 x}{\tan^2 x}$

(*g*) $\dfrac{\tan A + 1}{\tan A - 1} = \dfrac{1 + \cot A}{1 - \cot A}$

(*h*) $\cos A + \dfrac{\tan A}{\sec A} = \cos A + \sin A$

(*i*) $\dfrac{\cos y \sin y - \cos y}{\sin y} = \cot y(\sin y - 1)$

Answers

Exercise 8-3.1 **1** (*a*) sin *t* (*b*) 1 (*c*) 1

(*d*) tan *x* (*e*) $\dfrac{\tan^2 x}{\cot^2 x}$ (*f*) 2

2 $\sin x = \pm\sqrt{1 - \cos^2 x}$
3 $\cos x = \pm\sqrt{1 - \sin^2 x}$
4 $\tan A = \pm\sqrt{\sec^2 A - 1}$
5 $\tan 30° = \dfrac{1}{\sqrt{3}}, \dfrac{\sin 30°}{\cos 30°} = \dfrac{\frac{1}{2}}{\sqrt{3}/2} = \dfrac{1}{\sqrt{3}}$

$\tan 60° = \sqrt{3}, \dfrac{\sin 60°}{\cos 60°} = \dfrac{\sqrt{3}/2}{\frac{1}{2}} = \sqrt{3}$

Exercise 8-4.1 **1** *Proof*: $\dfrac{\sin x}{\csc x} + \dfrac{\cos x}{\sec x} = \dfrac{\sin x}{\frac{1}{\sin x}} + \dfrac{\cos x}{\frac{1}{\cos x}}$

$= \sin^2 x + \cos^2 x$
$= 1$

2 *Proof*: $\sec^2 A(1 - \sin^2 A) = \sec^2 A \cos^2 A$
$= 1$

3 *Proof*: $\dfrac{\cos B}{\sin B} + \dfrac{1 - \cos B}{\sin B} = \dfrac{1}{\sin B}$
$= \csc B$

4 *Proof*: $\dfrac{\sin C}{1 + \cos C} + \dfrac{1 + \cos C}{\sin C} = \dfrac{\sin^2 C + (1 + \cos C)^2}{(1 + \cos C)\sin C}$

$= \dfrac{\sin^2 C + 1 + 2 \cos C + \cos^2 C}{(1 + \cos C)\sin C}$

$= \dfrac{2 + 2 \cos C}{(1 + \cos C)\sin C}$

$= \dfrac{2(1 + \cos C)}{\sin C(1 + \cos C)} = \dfrac{2}{\sin C}$

$= 2 \csc C$

5 *Proof*: $\dfrac{\cos y(1 + \cos y) + \sin^2 y}{\sin y(1 + \cos y)} = \dfrac{\cos y + \cos^2 y + \sin^2 y}{\sin y(1 + \cos y)}$

$= \dfrac{\cos y + 1}{\sin y(1 + \cos y)} = \dfrac{1}{\sin y}$

$= \csc y$

6 *Proof*: $(\tan^3 A + \cot^3 A) = (\tan A + \cot A)(\tan^2 A - \tan A \cot A + \cot^2 A)$ by the sum-of-two-cubes factoring pattern.

7 *Proof*: $(\csc x - \sec x)(\csc x + \sec x) = \csc^2 x - \sec^2 x$

$$= \frac{1}{\sin^2 x} - \frac{1}{\cos^2 x}$$

$$= \frac{\cos^2 x - \sin^2 x}{\sin^2 x \cos^2 x}$$

8 *Proof*: $\dfrac{1 + \tan^2 x}{\csc^2 x} = \dfrac{\sec^2 x}{\csc^2 x}$

$$= \frac{\dfrac{1}{\cos^2 x}}{\dfrac{1}{\sin^2 x}}$$

$$= \frac{\sin^2 x}{\cos^2 x}$$

$$= \tan^2 x$$

9 *Proof*: $\dfrac{\sin B + 2 \sin B \cos B}{2 + \cos B - 2 \sin^2 B} = \dfrac{\sin B(1 + 2 \cos B)}{\cos B + 2 \cos^2 B}$

$$= \frac{\sin B(1 + 2 \cos B)}{\cos B(1 + 2 \cos B)}$$

$$= \frac{\sin B}{\cos B}$$

$$= \tan B$$

10 *Proof*: $\dfrac{\cot y + 2 \cos y}{\csc y - \sin y} = \dfrac{\dfrac{\cos y}{\sin y} + 2 \cos y}{\dfrac{1}{\sin y} - \sin y}$

$$= \frac{\cos y + 2 \cos y \sin y}{1 - \sin^2 y}$$

$$= \frac{\cos y(1 + 2 \sin y)}{\cos^2 y}$$

$$= \frac{1 + 2 \sin y}{\cos y}$$

$$= \frac{1}{\cos y} + \frac{2 \sin y}{\cos y}$$

$$= \sec y + 2 \tan y$$

11 *Proof*:
$$\frac{1 + \cot^2 z}{\sec^2 z} = \frac{\csc^2 z}{\sec^2 z}$$
$$= \csc^2 z \cdot \frac{1}{\sec^2 z}$$
$$= \csc^2 z \cdot \cos^2 z$$
$$= \frac{1}{\sin^2 z} \cdot \cos^2 z$$
$$= \frac{\cos^2 z}{\sin^2 z}$$
$$= \cot^2 z$$

12 *Proof*:
$$1 - \sin^4 x + \cos^4 x = (1 - \sin^2 x)(1 + \sin^2 x) + \cos^4 x$$
$$= \cos^2 x(1 + \sin^2 x) + \cos^4 x$$
$$= \cos^2 x(1 + \sin^2 x + \cos^2 x)$$
$$= \cos^2 x(1 + 1)$$
$$= 2\cos^2 x$$

13 *Proof*:
$$\frac{\sin A}{\csc A(1 + \cot^2 A)} = \frac{\sin A}{\dfrac{1}{\sin A}(\csc^2 A)}$$
$$= \frac{\sin A}{\dfrac{1}{\sin A}\dfrac{1}{\sin^2 A}} = \frac{\sin A}{\dfrac{1}{\sin^3 A}}$$
$$= \sin^4 A$$

14 *Proof*:
$$\sin^3 x + \cos^3 x = (\sin x + \cos x)(\sin^2 x - \sin x \cos x + \cos^2 x)$$
$$= (\sin x + \cos x)(1 - \sin x \cos x)$$

15 *Proof*:
$$\tan^2 x - \sin^2 x = \frac{\sin^2 x}{\cos^2 x} - \sin^2 x$$
$$= \frac{\sin^2 x - \sin^2 x \cos^2 x}{\cos^2 x}$$
$$= \frac{\sin^2 x(1 - \cos^2 x)}{\cos^2 x}$$
$$= \frac{\sin^2 x \cdot \sin^2 x}{\cos^2 x}$$
$$= \tan^2 x \sin^2 x$$

16 *Proof*:
$$\tan x + \cot x = \frac{\sin x}{\cos x} + \frac{\cos x}{\sin x}$$
$$= \frac{\sin^2 x + \cos^2 x}{\cos x \sin x}$$
$$= \frac{1}{\cos x \sin x}$$

17 *Proof:* $\dfrac{1 - \sec y}{1 + \sec y} = \dfrac{1 - \dfrac{1}{\cos y}}{1 + \dfrac{1}{\cos y}} = \dfrac{\dfrac{\cos y - 1}{\cos y}}{\dfrac{\cos y + 1}{\cos y}}$

$$= \dfrac{\cos y - 1}{\cos y + 1}$$

18 *Proof:* $\dfrac{1}{1 + \sin A} + \dfrac{1}{1 - \sin A} = \dfrac{1 - \sin A + 1 + \sin A}{1 - \sin^2 A}$

$$= \dfrac{2}{\cos^2 A}$$
$$= 2 \sec^2 A$$

19 *Proof:* $(\cos t - \sin t)(\sec t + \csc t) = (\cos t - \sin t)\left(\dfrac{1}{\cos t} + \dfrac{1}{\sin t}\right)$

$$= 1 - \tan t + \cot t - 1$$
$$= -\tan t + \cot t$$
$$= \cot t - \tan t$$

20 *Proof:* $\dfrac{\sin x}{\csc x} + \dfrac{\cos x}{\sec x} = \sin^2 x + \cos^2 x$

$$= 1$$
$$= \csc x \sin x$$

21 *Proof:* $\dfrac{\sin^4 x - \cos^4 x}{\sin^2 x - \cos^2 x} = \dfrac{(\sin^2 x - \cos^2 x)(\sin^2 x + \cos^2 x)}{\sin^2 x - \cos^2 x}$
$$= \sin^2 x + \cos^2 x$$
$$= 1$$

22 *Proof:* $\dfrac{1 + \sec^3 m}{1 + \sec m} = \dfrac{(1 + \sec m)(1 - \sec m + \sec^2 m)}{1 + \sec m}$
$$= 1 - \sec m + \sec^2 m$$

23 *Proof:* $\sin^2 x + \csc^2 x + \cos^2 x - \cot^2 x + \sec^2 x - \tan^2 x$
$$= (\sin^2 x + \cos^2 x) + (\csc^2 x - \cot^2 x) + (\sec^2 x - \tan^2 x)$$
$$= 1 + 1 + 1$$
$$= 3$$

Review Exercise **1** (*a*) 1 (*b*) 1 (*c*) 1 (*d*) $\tan x$
(*e*) $\cot x$ (*f*) $\sec^2 x$ (*g*) $\csc^2 x$ (*h*) 1
2 (*a*) $6 \tan^3 x$ (*b*) $\cos^2 x$ (*c*) $1 + 2 \tan^2 z$
(*d*) $\sin^2 y$ (*e*) 0 (*f*) $\csc^2 x$

3 (a) *Proof*: $\dfrac{\csc^2 A - \cot^2 A}{\sec^2 A} = \dfrac{(\cot^2 A + 1) - \cot^2 A}{\sec^2 A}$

$$= \dfrac{1}{\sec^2 A}$$
$$= \cos^2 A$$

(b) *Proof*: $\dfrac{\cos^2 x \sin^2 x + \cos^4 x}{\sin^2 x} = \dfrac{\cos^2 x(\sin^2 x + \cos^2 x)}{\sin^2 x}$

$$= \dfrac{\cos^2 x \cdot 1}{\sin^2 x}$$
$$= \cot^2 x$$

(c) *Proof*: $\cot B \cos B = \dfrac{\cos B}{\sin B} \cdot \cos B$

$$= \dfrac{\cos^2 B}{\sin B}$$
$$= \dfrac{1 - \sin^2 B}{\sin B}$$
$$= \dfrac{1}{\sin B} - \dfrac{\sin^2 B}{\sin B}$$
$$= \csc B - \sin B$$

(d) *Proof*: $\dfrac{\cos^4 C - \sin^4 C}{\cos C + \sin C} = \dfrac{(\cos^2 C - \sin^2 C)(\cos^2 C + \sin^2 C)}{\cos C + \sin C}$

$$= \dfrac{\cos^2 C - \sin^2 C(1)}{\cos C + \sin C}$$
$$= \dfrac{(\cos C - \sin C)(\cos C + \sin C)}{\cos C + \sin C}$$
$$= \cos C - \sin C$$

(e) *Proof*: $\dfrac{\cos^2 x - \sin^2 x \cos^2 x}{\sin^2 x} = \dfrac{\cos^2 x(1 - \sin^2 x)}{\sin^2 x}$

$$= \dfrac{\cos^2 x \cdot \cos^2 x}{\sin^2 x}$$
$$= \cot^2 x \cos^2 x$$

(f) *Proof*:
$$\tan^2 x(1 - \sin^2 x) + \cot^2 x = \tan^2 x \cos^2 x + \cot^2 x$$
$$= \frac{\sin^2 x}{\cos^2 x} \cdot \cos^2 x + \frac{\cos^2 x}{\sin^2 x}$$
$$= \sin^2 x + \frac{\cos^2 x}{\sin^2 x}$$
$$= \frac{\sin^4 x + \cos^2 x}{\sin^2 x} = \frac{\sin^4 x + (1 - \sin^2 x)}{\sin^2 x}$$
$$= \frac{1 - 2\sin^2 x + \sin^2 x + \sin^4 x}{\sin^2 x}$$
$$= \frac{\sin^2 x + (1 - \sin^2 x)^2}{\sin^2 x} = \frac{\sin^2 x}{\sin^2 x} + \frac{\cos^4 x}{\sin^2 x}$$
$$= 1 + \frac{\cos^2 x}{\dfrac{\sin^2 x}{\cos^2 x}} = 1 + \frac{\cos^2 x}{\tan^2 x}$$

(g) *Proof*:
$$\frac{\tan A + 1}{\tan A - 1} = \frac{\dfrac{1}{\cot A} + 1}{\dfrac{1}{\cot A} - 1}$$
$$= \frac{\dfrac{1 + \cot A}{\cot A}}{\dfrac{1 - \cot A}{\cot A}}$$
$$= \frac{1 + \cot A}{1 - \cot A}$$

(h) *Proof*:
$$\cos A + \frac{\tan A}{\sec A} = \cos A + \frac{\dfrac{\sin A}{\cos A}}{\dfrac{1}{\cos A}}$$
$$= \cos A + \sin A$$

(i) *Proof*:
$$\frac{\cos y \sin y - \cos y}{\sin y} = \frac{\cos y(\sin y - 1)}{\sin y}$$
$$= \cot y(\sin y - 1)$$

Additional Identities

9

Objectives

Upon completion of Chapter 9, you will be able to:

1 Write from memory the following identities:

$$\sin(-\theta) = -\sin\theta \qquad (9\text{-}3)$$

$$\sin\left(\frac{\pi}{2} - \theta\right) = \cos\theta \qquad (9\text{-}2)$$

$$\cos(-\theta) = \cos\theta \qquad (9\text{-}3)$$

$$\cos\left(\frac{\pi}{2} - \theta\right) = \sin\theta \qquad (9\text{-}2)$$

$$\cos(A + B) = \\ \cos A \cos B - \sin A \sin B \qquad (9\text{-}4)$$

$$\cos(A - B) = \\ \cos A \cos B + \sin A \sin B \qquad (9\text{-}1)$$

$$\sin(A + B) = \\ \sin A \cos B + \cos A \sin B \qquad (9\text{-}4)$$

$$\sin(A - B) = \\ \sin A \cos B - \cos A \sin B \qquad (9\text{-}4)$$

$$\tan(A + B) = \frac{\tan A + \tan B}{1 - \tan A \tan B} \qquad (9\text{-}4)$$

$$\tan(A - B) = \frac{\tan A - \tan B}{1 + \tan A \tan B} \qquad (9\text{-}4)$$

$$\sin 2A = 2 \sin A \cos A \qquad (9\text{-}5)$$

$$\cos 2A = \cos^2 A - \sin^2 A \qquad (9\text{-}5)$$

$$\cos 2A = 2 \cos^2 A - 1 \qquad (9\text{-}5)$$

$$\cos 2A = 1 - 2 \sin^2 A \qquad (9\text{-}5)$$

$$\tan 2A = \frac{2 \tan A}{1 - \tan^2 A} \qquad (9\text{-}5)$$

$$\sin \frac{1}{2}A = \pm\sqrt{\frac{1 - \cos A}{2}} \qquad (9\text{-}6)$$

$$\cos \frac{1}{2}A = \pm\sqrt{\frac{1 + \cos A}{2}} \qquad (9\text{-}6)$$

$$\tan \frac{A}{2} = \pm\sqrt{\frac{1 - \cos A}{1 + \cos A}} \qquad (9\text{-}6)$$

$$\tan \frac{A}{2} = \frac{\sin A}{1 + \cos A} \qquad (9\text{-}6)$$

$$\tan \frac{A}{2} = \frac{1 - \cos A}{\sin A} \qquad (9\text{-}6)$$

2 Write the cofunction of any given trigonometric function of any angle. (9-2)

3 Simplify expressions by using the identities in objective 1. (9-3, 9-4)

4 Obtain the exact value of $T(A + B)$ where T is a trigonometric function, A and B are angles. (9-4)

5 Prove identities by using the identities in objective 1. (9-4 to 9-6)

9/Additional Identities

9-1 Cosine of Difference of Two Angles

There are many instances when trigonometric functions of more than one angle are used.   One formula containing two angles is the identity:

$$\cos (A - B) = \cos B \cos B + \sin A \sin B$$

A proof for this identity is given below.

Figure 9-1.1

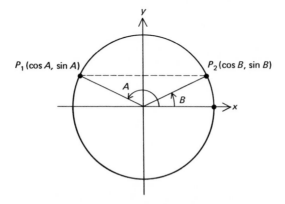

Consider the unit circle with angles A and B in standard position (Figure 9-1.1).   If A is an angle and P_1 is a point on the terminal side of A and also on the unit circle, then the x coordinate of A is $\cos A$ and the y coordinate is $\sin A$.   Similarly, $P_2(\cos B, \sin B)$ can be considered as a point on the terminal side of B and also on the unit circle.   Now, using the distance formula, we get

$$\overline{P_1P_2}^2 = (\cos A - \cos B)^2 + (\sin A - \sin B)^2$$

Next, rotate angles A and B clockwise until P_2 coincides with $(1,0)$ and the terminal side of B is on the positive x axis, as in Figure 9-1.2.

With this rotation, angle $A - B$ is placed in standard position. Hence,

$$\overline{P_1P_2}^2 = [\cos (A - B) - 1]^2 + [\sin (A - B) - 0]^2$$

Thus,

$$[\cos (A - B) - 1]^2 + [\sin (A - B) - 0]^2 = (\cos A - \cos B)^2 + (\sin A - \sin B)^2$$

$$\cos^2 (A - B) - 2 \cos (A - B) + 1 + \sin^2 (A - B) = \cos^2 A - 2 \cos A \cos B + \cos^2 B + \sin^2 A - 2 \sin A \sin B + \sin^2 B$$

Figure 9-1.2

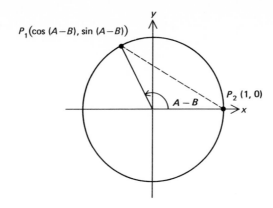

Using $\sin^2 A + \cos^2 A = 1$, we have

$$2 - 2 \cos (A - B) = 2 - 2 \cos A \cos B - 2 \sin A \sin B$$

$$-2 \cos (A - B) = -2(\cos A \cos B + \sin A \sin B)$$

and finally,

$$\cos (A - B) = \cos A \cos B + \sin A \sin B$$

It is a good idea to learn what the identity says in words: the cosine of the difference of two angles, $A - B$, is the cosine of the first times the cosine of the second plus the sine of the first times the sine of the second. Note also that $\cos (A - B) \neq \cos A - \cos B$.

9-2 Cofunction Identities

You may have guessed during your study of Chapter 2 that $\sin 60° = \cos 30°$, $\tan 75° = \cot 15°$, and $\sec 40° = \csc 50°$. These are examples of cofunction identities. We are now able to prove the cofunction identities by using the formula for $\cos (A - B)$.

$$\cos \left(\frac{\pi}{2} - A\right) = \cos \frac{\pi}{2} \cos A + \sin \frac{\pi}{2} \sin A$$

$$= 0 \cdot \cos A + 1 \cdot \sin A$$

$$= \sin A$$

Hence,

$$\cos \left(\frac{\pi}{2} - A\right) = \sin A$$

Next we obtain a similar formula for $\sin \left(\frac{\pi}{2} - A\right)$.

$$\sin\left(\frac{\pi}{2} - A\right) = \cos\left[\frac{\pi}{2} - \left(\frac{\pi}{2} - A\right)\right] \quad \text{from the previous identity}$$

$$= \cos\left(\frac{\pi}{2} - \frac{\pi}{2} + A\right)$$

$$\sin\left(\frac{\pi}{2} - A\right) = \cos A$$

Exercise 9-2.1 **1** Express each function as its cofunction; for example, if $\sin 305° = \sin (90° - A)$, then $A = -215°$; so $\sin 305° = \cos (-215°)$.

(*a*) $\cos 32°$ (*b*) $\sin 18°37'$ (*c*) $\cos (-10°)$

(*d*) $\sin \dfrac{2\pi}{3}$ (*e*) $\cos 113°45'$ (*f*) $\cos 10$

9-3 Negative of an Angle If we let $A = 0°$ in the formula for $\cos (A - B)$, we have

$$\cos (-B) = \cos (0° - B)$$

$$= \cos 0° \cos B + \sin 0° \sin B$$

$$= 1(\cos B) + 0(\sin B)$$

$$= \cos B$$

therefore,

$$\cos (-B) = \cos B$$

Using $\cos\left(\dfrac{\pi}{2} - A\right) = \sin A$, we can write

$$\sin (-B) = \cos\left[\frac{\pi}{2} - (-B)\right]$$

$$= \cos\left(\frac{\pi}{2} + B\right)$$

$$= \cos\left[-\left(\frac{\pi}{2} + B\right)\right] \quad \text{from the previous identity}$$

$$= \cos \left[-\left(\frac{\pi}{2}\right) - B \right]$$

$$= \cos \left(-\frac{\pi}{2}\right) \cos B + \sin \left(-\frac{\pi}{2}\right) \sin B$$

$$= 0 \cdot \cos B + (-1) \sin B$$

$$= -\sin B$$

Therefore,

$$\sin (-B) = -\sin B$$

The following examples illustrate the identities of this section.

Example 9-3.1 Write $\cos 25°$ as the same function of $(-25°)$.

Solution $\cos 25° = \cos (-25°)$

Example 9-3.2 Write $\sin 38°$ as the same function of $(-38°)$.

Solution $\sin 38° = -\sin (-38°)$

Example 9-3.3 Simplify $\dfrac{-\sin (-x)}{\cos (-x)}$

Solution $\dfrac{-\sin (-x)}{\cos (-x)} = \dfrac{\sin x}{\cos x}$

$$= \tan x$$

Exercise 9-3.1 **1** Use $\sin (-x) = -\sin x$ and $\cos (-x) = \cos x$ to write each given function of an angle as the same function of the negative of the angle:

(a) $\sin (-50°)$ (b) $\cos (-50°)$ (c) $\sin 50°$

(d) $\cos 100°$ (e) $\sin (-\sqrt{3})$ (f) $\cos 13°$

2 Simplify $\dfrac{-\cos x}{\sin (-x)}$

9-4 Additional Identities for Sum and Difference of Angles

The previous five identities can be used to prove additional identities that contain more than one angle. They are listed below, and the proof of each is left as an exercise for the student.

$$\cos (A + B) = \cos A \cos B - \sin A \sin B$$

$$\sin (A + B) = \sin A \cos B + \cos A \sin B$$

$$\sin (A - B) = \sin A \cos B - \cos A \sin B$$

$$\tan (A + B) = \frac{\tan A + \tan B}{1 - \tan A \tan B}$$

$$\tan (A - B) = \frac{\tan A - \tan B}{1 + \tan A \tan B}$$

(for all values where tan A, tan B are defined and the denominator is not zero)

The following examples make use of the identities for sums and differences of angles.

Example 9-4.1 Evaluate sin 15°.

Solution

$$\sin 15° = \sin (45° - 30°)$$

$$= \sin 45° \cos 30° - \cos 45° \sin 30°$$

$$= \frac{\sqrt{2}}{2} \cdot \frac{\sqrt{3}}{2} - \frac{\sqrt{2}}{2} \cdot \frac{1}{2}$$

$$= \frac{\sqrt{6}}{4} - \frac{\sqrt{2}}{4}$$

$$= \frac{\sqrt{6} - \sqrt{2}}{4}$$

Example 9-4.2 Evaluate $\tan \dfrac{\pi}{12}$.

Solution

$$\tan \frac{\pi}{12} = \tan \left(\frac{4\pi}{12} - \frac{3\pi}{12} \right) = \tan \left(\frac{\pi}{3} - \frac{\pi}{4} \right)$$

$$= \frac{\tan \dfrac{\pi}{3} - \tan \dfrac{\pi}{4}}{1 + \tan \dfrac{\pi}{3} \tan \dfrac{\pi}{4}}$$

$$= \frac{\sqrt{3} - 1}{1 + \sqrt{3}(1)}$$

$$= \frac{\sqrt{3} - 1}{1 + \sqrt{3}} \cdot \frac{1 - \sqrt{3}}{1 - \sqrt{3}}$$

$$= \frac{\sqrt{3} - 3 - 1 + \sqrt{3}}{1 - 3}$$

$$= \frac{2\sqrt{3} - 4}{-2}$$

$$= -\sqrt{3} + 2$$

Example 9-4.3 Simplify the expression $\cos \left(\frac{3\pi}{2} + A \right)$.

Solution
$$\cos \left(\frac{3\pi}{2} + A \right) = \cos \frac{3\pi}{2} \cos A - \sin \frac{3\pi}{2} \sin A$$

$$= 0 \cdot \cos A - (-1) \sin A$$

$$= \sin A$$

Exercise 9-4.1 **1** Prove each of the identities listed in Section 9-4.

2 Simplify each of the following expressions:

(a) $\dfrac{\tan 6C + \tan (-3C)}{1 - \tan 6C \tan (-3C)}$

(b) $\sin \left(\dfrac{\pi}{2} - 2x \right)$

(c) $\cos \left(\dfrac{\pi}{2} - 2x \right)$

(d) $\cos 5 \cos 10 - \sin 5 \sin 10$

(e) $\cos (x + y) + \cos (x - y)$

3 Prove each of the following identities:

(a) $\sin(\pi - A) = \sin A$

(b) $\cos(\pi - A) = -\cos A$

(c) $\tan\left(\dfrac{\pi}{2} - A\right) = \cot A$

(d) $\sec\left(\dfrac{\pi}{2} - t\right) = \csc t$

(e) $\dfrac{\tan(x + y) - \tan y}{1 + \tan(x + y)\tan y} = \tan x$

(f) $\dfrac{\tan(x + y)}{\cot(x - y)} = \dfrac{\tan^2 x - \tan^2 y}{1 - \tan^2 x \tan^2 y}$

(g) $\tan(225° - x) = \dfrac{1 - \tan x}{1 + \tan x}$

(h) $\cot(x + y) = \dfrac{\cot x \cot y - 1}{\cot x + \cot y}$

(i) $\sin(A + B) - \sin(A - B) = 2\cos A \sin B$

4 Find the exact value of each of the following:

(a) $\tan(A - B)$, if $\sec A = \frac{13}{5}$, $\csc B = \frac{17}{8}$, where A is in Q I and B is in Q II.

(b) cos 75° (**Hint:**   75° = 135° − 60°.)

(c) sin $(k + t)$, if $0 < k < \dfrac{\pi}{2}$, sin $k = \dfrac{4}{5}$, $\pi < t < \dfrac{3\pi}{2}$, and cos $t = -\dfrac{3}{5}$.

(d) cos $(k + t)$, same conditions as problem 4c.

(e) tan $(k - t)$, same conditions as problem 4c.

9-5 Double-Angle Identities

By letting $A = B$ in the formulas for the sine, cosine, and tangent of $A + B$, we obtain the identities for the functions of twice an angle, i.e., the double-angle formulas. For example,

$$\sin 2A = \sin (A + A)$$

$$= \sin A \cos A + \cos A \sin A$$

and

$$\sin 2A = 2 \sin A \cos A$$

Also,

$$\cos 2A = \cos (A + A)$$

$$= \cos A \cos A - \sin A \sin A$$

hence

$$\cos 2A = \cos^2 A - \sin^2 A$$

$$\cos 2A = 2 \cos^2 A - 1$$

and

$$\cos 2A = 1 - 2 \sin^2 A$$

Similarly,

$$\tan 2A = \frac{2 \tan A}{1 - \tan^2 A}$$

Example 9-5.1 Find $\sin 2B$, $\cos 2B$, and $\tan 2B$, if $\sin B = -\dfrac{3}{5}$ and $\dfrac{3\pi}{2} < B < 2\pi$.

Solution $x = 4$ $r = 5$ $y = -3$

Figure 9-5.1

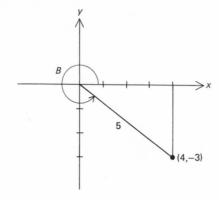

$$\sin 2B = 2 \sin B \cos B$$

$$= 2\left(-\frac{3}{5}\right)\left(\frac{4}{5}\right)$$

$$= -\frac{24}{25}$$

$$\cos 2B = \cos^2 B - \sin^2 B$$

$$= \left(\frac{4}{5}\right)^2 - \left(-\frac{3}{5}\right)^2$$

$$= \frac{16}{25} - \frac{9}{25} = \frac{7}{25}$$

$$\tan 2B = \frac{2 \tan B}{1 - \tan^2 B}$$

$$= \frac{2\left(-\frac{3}{4}\right)}{1 - \left(-\frac{3}{4}\right)^2}$$

$$= \frac{-\frac{3}{2}}{1 - \frac{9}{16}} = \left(-\frac{3}{2}\right)\left(\frac{16}{7}\right) = -\frac{24}{7}$$

Example 9-5.2   Evaluate $\cos \dfrac{2\pi}{3}$ by using a double-angle formula.

Solution
$$\cos \frac{2\pi}{3} = \cos 2\left(\frac{\pi}{3}\right)$$

$$= \cos^2 \frac{\pi}{3} - \sin^2 \frac{\pi}{3}$$

$$= \left(\frac{1}{2}\right)^2 - \left(\frac{\sqrt{3}}{2}\right)^2$$

$$= \frac{1}{4} - \frac{3}{4} = -\frac{1}{2}$$

Exercise 9-5.1   **1**   Find $\sin 2A$, $\cos 2A$, and $\tan 2A$ in each problem.

(a) $\cos A = -\frac{12}{13}$, A in Q II

(b) $\tan A = \frac{12}{5}$, $\pi < A < \dfrac{3\pi}{2}$

(c) $\csc A = -\frac{17}{15}$, A in Q III

(d) $\cot A = r$, $0 < A < \dfrac{\pi}{2}$

2   Use a double-angle formula to give the exact value of $\tan 120°$.

3   Prove each of the following identities:

(a) $\sin 2x = \dfrac{2 \cos^2 x}{\cot x}$

(b) $\tan x \sin 2x = 2 \sin^2 x$

(c) $2 \cot 2A = \cot A - \tan A$ (d) $\dfrac{\cos 2A}{\cos A - \sin A} = \cos A + \sin A$

(e) $\cos 10A = 1 - 2 \sin^2 5A$ (f) $\cos^4 t - \sin^4 t = \cos 2t$

(g) $\cos^3 t - \sin^3 t = (\cos t - \sin t)(1 + \tfrac{1}{2} \sin 2t)$

(h) $\dfrac{1 - \cos 2x}{2} = \sin^2 x$

(i) $\cos^6 B - \sin^6 B = \cos 2B \, (1 - \sin^2 B \cos^2 B)$

(j) $\cos^4 x - \sin^4 x = 2 \cos^2 x - 1$

9-6 Half-Angle Identities By using two double-angle identities we can obtain formulas for $\sin \tfrac{1}{2}x$ and $\cos \tfrac{1}{2}x$. They are used frequently in calculus and other courses. Since $\cos 2x = 1 - 2 \sin^2 x$,

$$\cos x = 1 - 2 \sin^2 \tfrac{1}{2}x$$

Hence,

$$2 \sin^2 \tfrac{1}{2}x = 1 - \cos x$$

$$\sin^2 \tfrac{1}{2}x = \frac{1 - \cos x}{2}$$

and

$$\sin \tfrac{1}{2}x = \pm \sqrt{\frac{1 - \cos x}{2}}$$

Similarly,

$$\cos 2x = 2 \cos^2 x - 1$$

$$2 \cos^2 x = \cos 2x + 1$$

$$2 \cos^2 \tfrac{1}{2}x = \cos x + 1$$

and

$$\cos \tfrac{1}{2}x = \pm \sqrt{\frac{1 + \cos x}{2}}$$

To obtain a formula for $\tan \tfrac{1}{2}x$, recall that $\tan \tfrac{1}{2}x = \dfrac{\sin \tfrac{1}{2}x}{\cos \tfrac{1}{2}x}$. Hence,

$$\tan \tfrac{1}{2}x = \frac{\pm \sqrt{\dfrac{1 - \cos x}{2}}}{\pm \sqrt{\dfrac{1 + \cos x}{2}}}$$

and

$$\tan \tfrac{1}{2}x = \pm \sqrt{\frac{1 - \cos x}{1 + \cos x}}$$

The plus or minus sign is chosen according to the quadrant in which $\dfrac{x}{2}$ lies. Two other forms often used for $\tan \tfrac{1}{2}x$ are

$$\tan \tfrac{1}{2}x = \frac{\sin x}{1 + \cos x}$$

and

$$\tan \tfrac{1}{2}x = \frac{1 - \cos x}{\sin x}$$

The first of these is obtained by multiplying the numerator and denominator of $\pm \sqrt{\dfrac{1 - \cos x}{1 + \cos x}}$ by $(1 + \cos x)$. The second is proven by multiplying the numerator and denominator by $(1 - \cos x)$. The reader may wish to verify these.

Example 9-6.1 Find the exact value of $\sin \dfrac{A}{2}$, $\cos \dfrac{A}{2}$, and $\tan \dfrac{A}{2}$ if $\cot A = \dfrac{4}{3}$ and $\pi < A < \dfrac{3\pi}{2}$.

Solution Sketch A in standard position. Then $x = -4$, $r = 5$, and $y = -3$.

Figure 9-6.1

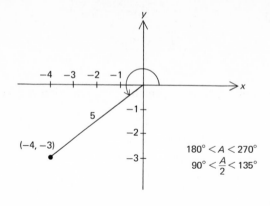

$$180° < A < 270°$$
$$90° < \frac{A}{2} < 135°$$

$$\sin \frac{A}{2} = \sqrt{\frac{1 - \cos A}{2}} \qquad \text{since } \frac{A}{2} \text{ is in Q II}$$

$$= \sqrt{\frac{1 - (-\frac{4}{5})}{2}}$$

$$= \sqrt{\frac{9}{10}} = \frac{3}{\sqrt{10}} \text{ or } \frac{3\sqrt{10}}{10}$$

$$\cos \frac{A}{2} = -\sqrt{\frac{1 + \cos A}{2}} = -\sqrt{\frac{1 - \frac{4}{5}}{2}}$$

$$= -\sqrt{\frac{1}{10}} = -\frac{\sqrt{10}}{10}$$

$$\tan \frac{A}{2} = \frac{\sin A}{1 + \cos A} = \frac{-\frac{3}{5}}{1 - \frac{4}{5}} = -3$$

Example 9-6.2 A problem often encountered in calculus requires the following task: Express $\cos^4 x$ as a function of only first powers of $\cos nx$, n is a natural number.

Solution Since

$$\cos x = \pm\sqrt{\frac{1 + \cos 2x}{2}}$$

$$\cos^2 x = \frac{1 + \cos 2x}{2}$$

and

$$\cos^4 x = \frac{1 + 2\cos 2x + \cos^2 2x}{4}$$

$$= \frac{1 + 2\cos 2x + \dfrac{1 + \cos 4x}{2}}{4}$$

$$= \frac{2 + 4\cos 2x + 1 + \cos 4x}{8}$$

$$= \frac{3 + 4\cos 2x + \cos 4x}{8}$$

Example 9-6.3 Use a half-angle formula to find the value of $\sin 15°$.

Solution $\sin 15° = \sin \tfrac{1}{2}(30°) = \sqrt{\dfrac{1 - \cos 30°}{2}}$

$$= \sqrt{\dfrac{1 - (\sqrt{3}/2)}{2}}$$

$$= \tfrac{1}{2}\sqrt{2 - \sqrt{3}}$$

Exercise 9-6.1 **1** In each of the following, find $\sin \tfrac{1}{2}A$, $\cos \tfrac{1}{2}A$, and $\tan \tfrac{1}{2}A$.

(a) $\sin A = \tfrac{4}{5}$, $0° < A < 90°$

(b) $\csc A = -\tfrac{17}{15}$, $180° < A < 270°$

(c) $\sec A = -\tfrac{17}{8}$, $\dfrac{\pi}{2} < A < \pi$

2 Use half-angle formulas to evaluate each of the following:

(a) $\cos 15°$ (b) $\tan 22.5°$

(c) $\cot 75°$ (d) $\sin \dfrac{\pi}{8}$

(e) sec 105°   (f) sin 112.5°

(g) cos 157.5°   (h) $\cos \dfrac{5\pi}{12}$

3 Verify the following identities:

(a) $(1 + \cos 2x)\left(1 + \tan \dfrac{x}{2} \tan x\right)^2 = 2$

(b) $\sin \dfrac{x}{2} \cos \dfrac{x}{2} = \dfrac{1}{2} \sin x$

(c) $\sec^2 \dfrac{x}{2} = \dfrac{2 \sec x}{1 + \sec x}$

(d) $\cos A - \sin \dfrac{A}{2} = \left(1 - 2 \sin \dfrac{A}{2}\right)\left(1 + \sin \dfrac{A}{2}\right)$

(e) $2 \cot A = \cot \dfrac{A}{2} - \tan \dfrac{A}{2}$

(f) $1 + \cos A = 2 \cos^2 \dfrac{A}{2}$

(g) $\sec \dfrac{x}{2} + \csc \dfrac{x}{2} = 2 \csc^2 x\left(\sin \dfrac{x}{2} + \cos \dfrac{x}{2}\right)$

(h) $\cot \dfrac{B}{2} = \dfrac{\sin B}{1 - \cos B}$

$$(i) \quad \cot^2 \frac{B}{2} = \frac{1 + \cos B}{1 - \cos B}$$

$$(j) \quad 2 \sin^2 A = 8 \sin^2 \tfrac{1}{2}A \cos^2 \tfrac{1}{2}A$$

9-7 Summary of Additional Identities

So that you may have a list of all the major identities given in this chapter we present them here for easy referral.

$$\sin (-\theta) = -\sin \theta$$

$$\cos (-\theta) = \cos \theta$$

$$\sin \left(\frac{\pi}{2} - \theta \right) = \cos \theta$$

$$\cos \left(\frac{\pi}{2} - \theta \right) = \sin \theta$$

$$\cos (A + B) = \cos A \cos B - \sin A \sin B$$

$$\cos (A - B) = \cos A \cos B + \sin A \sin B$$

$$\sin (A + B) = \sin A \cos B + \cos A \sin B$$

$$\sin (A - B) = \sin A \cos B - \cos A \sin B$$

$$\tan (A - B) = \frac{\tan A - \tan B}{1 + \tan A \tan B} \qquad \tan (A + B) = \frac{\tan A + \tan B}{1 - \tan A \tan B}$$

$$\sin 2A = 2 \sin A \cos A$$

$$\cos 2A = \cos^2 A - \sin^2 A$$

$$\cos 2A = 1 - 2 \sin^2 A$$

$$\cos 2A = 2 \cos^2 A - 1$$

$$\tan 2A = \frac{2 \tan A}{1 - \tan^2 A}$$

$$\sin \tfrac{1}{2}A = \pm \sqrt{\frac{1 - \cos A}{2}}$$

$$\cos \tfrac{1}{2}A = \pm \sqrt{\frac{1 + \cos A}{2}}$$

$$\tan \tfrac{1}{2}A = \pm \sqrt{\frac{1 - \cos A}{1 + \cos A}}$$

$$\tan \tfrac{1}{2}A = \frac{\sin A}{1 + \cos A}$$

$$\tan \tfrac{1}{2}A = \frac{1 - \cos A}{\sin A}$$

Review Exercise   **1**   Fill in the blanks in each statement below.

(a) $\cos \left(\dfrac{\pi}{2} - \theta \right) =$ _____

(b) $\sin (A + B) = \sin A$ _____ + _____ $\sin B$

(c) $\tan (A - B) = \dfrac{\tan A + \text{_____}}{1 - \text{_____}}$

(d) $\cos 2A = 2 \cos^2 A -$ _____

(e) $\sin \tfrac{1}{2}x =$ _____

(f) $\tan \dfrac{C}{2} = \dfrac{\sin C}{1 + \text{_____}}$

2   Write the cofunction of each function given below.

(a) $\cos 73°37'$ (b) $\tan 21°13'$

(c) $\sin 200°$ (d) $\sec 110°$

(e) $\csc (-100°)$ (f) $\cot (-50°)$

(g) $\sin (-20°39')$ (h) $\cos 500°$

3 Prove each of the following identities:

(a) $\cot x \sin x = \cos x$

(b) $\sec A \tan A = \dfrac{\sin A}{\cos^2 A}$

(c) $\tan A \sin A + \cos A = \sec A$

(d) $\tan A + \dfrac{\cos A}{1 + \sin A} = \sec A$

(e) $\tan (45° + x) = \dfrac{1 + \tan x}{1 - \tan x}$

(f) $\dfrac{1 + \cos 2C}{1 - \sin C} = 2(1 + \sin C)$

(g) $\dfrac{\sec^2 \frac{1}{2}x}{\tan^2 \frac{1}{2}x} = 2 \csc^2 x(1 + \cos x)$

(h) $\frac{1}{2}\csc^2 x = \dfrac{2 \cos^2 x}{\sin^2 2x}$

(i) $\dfrac{4 \cos^4 \frac{1}{2}x}{\tan^2 x} = \dfrac{\cos^2 x}{\tan^2 \frac{1}{2}x}$

4 Write the exact value of each of the following:

(a) $\sin 157.5°$

(b) $\cos 112.5°$

(c) $\sin (A + B)$ if $\sin A = \frac{12}{13}$, $\cos B = \frac{7}{25}$, and A and B are in Q I

(d) $\tan (A + B)$ if $\tan A = -\frac{8}{15}$, $\tan B = \frac{5}{12}$, A in Q II, B in Q III

5 Simplify each expression:

(a) $\sin 3k \cos 2k - \cos 3k \sin 2k$ (b) $\dfrac{\tan 5M + \tan 3M}{1 - \tan 5M \tan 3M}$

(c) $\sqrt{\dfrac{1 - \cos 10°}{2}}$ (d) $\dfrac{\sin 35°}{1 + \cos 35°}$

(e) $2 \sin 45° \cos 45°$ (f) $(\sin x - \cos x)^2 + \sin 2x - 1$

Answers

Exercise 9-2.1 **1**   (a) $\cos 32° = \cos (90° - 58°) = \sin 58°$
(b) $\sin (18°37') = \sin (90° - 71°23') = \cos 71°23'$
(c) $\cos (-10°) = \cos (90° - 100°) = \sin 100° = \sin 80°$
(d) $\sin \dfrac{2\pi}{3} = \sin \left[\dfrac{\pi}{2} - \left(-\dfrac{\pi}{6} \right) \right] = \cos \left(-\dfrac{\pi}{6} \right) = \cos \dfrac{\pi}{6}$
(e) $\cos 113°45' = \cos [90° - (-23°45')] = \sin (-23°45') = -\sin 23°45'$
(f) $\cos 10 = \cos [1.57 - (-8.43)] = \sin (-8.43) = -\sin 1.15$

Exercise 9-3.1 **1**   (a) $\sin (-50°) = -\sin 50°$ (b) $\cos (-50°) = \cos 50°$
(c) $\sin 50° = -\sin (-50°)$ (d) $\cos 100° = \cos (-100°)$
(e) $\sin (-\sqrt{3}) = -\sin \sqrt{3}$ (f) $\cos 13° = \cos (-13°)$

2 $\dfrac{-\cos x}{\sin (-x)} = \dfrac{-\cos x}{-\sin x} = \dfrac{\cos x}{\sin x} = \cot x$

Exercise 9-4.1 **1** $\cos (A + B) = \cos [A - (-B)]$
$\qquad\qquad\qquad = \cos A \cos (-B) + \sin A \sin (-B)$
$\qquad\qquad\qquad = \cos A \cos B + \sin A (-\sin B)$
$\qquad\qquad\qquad = \cos A \cos B - \sin A \sin B$

$\sin (A + B) = \cos \left[\dfrac{\pi}{2} - (A + B) \right]$
$\qquad\qquad\quad = \cos \left[\left(\dfrac{\pi}{2} - A \right) - B \right]$
$\qquad\qquad\quad = \cos \left(\dfrac{\pi}{2} - A \right) \cos B + \sin \left(\dfrac{\pi}{2} - A \right) \sin B$
$\qquad\qquad\quad = \sin A \cos B + \cos A \sin B$

$$\sin (A - B) = \sin [A + (-B)]$$
$$= \sin A \cos (-B) + \cos A \sin (-B)$$
$$= \sin A \cos B + \cos A (-\sin B)$$
$$= \sin A \cos B - \cos A \sin B$$

$$\tan (A + B) = \frac{\sin (A + B)}{\cos (A + B)}$$
$$= \frac{\sin A \cos B + \cos A \sin B}{\cos A \cos B - \sin A \sin B}$$
$$= \frac{\dfrac{\sin A \cos B}{\cos A \cos B} + \dfrac{\cos A \sin B}{\cos A \cos B}}{\dfrac{\cos A \cos B}{\cos A \cos B} - \dfrac{\sin A \sin B}{\cos A \cos B}}$$
$$= \frac{\dfrac{\sin A}{\cos A} + \dfrac{\sin B}{\cos B}}{1 - \dfrac{\sin A}{\cos A} \cdot \dfrac{\sin B}{\cos B}} = \frac{\tan A + \tan B}{1 - \tan A \tan B}$$

$$\tan (A - B) = \frac{\sin (A - B)}{\cos (A - B)}$$
$$= \frac{\sin A \cos B - \cos A \sin B}{\cos A \cos B + \sin A \sin B}$$
$$= \frac{\dfrac{\sin A \cos B}{\cos A \cos B} - \dfrac{\cos A \sin B}{\cos A \cos B}}{\dfrac{\cos A \cos B}{\cos A \cos B} + \dfrac{\sin A \sin B}{\cos A \cos B}}$$
$$= \frac{\dfrac{\sin A}{\cos A} - \dfrac{\sin B}{\cos B}}{1 + \dfrac{\sin A}{\cos A} \cdot \dfrac{\sin B}{\cos B}} = \frac{\tan A - \tan B}{1 + \tan A \tan B}$$

2 (a) $\dfrac{\tan 6C + \tan (-3C)}{1 - \tan 6C \tan (-3C)} = \tan [6C + (-3C)] = \tan 3C$

(b) $\sin \left(\dfrac{\pi}{2} - 2x\right) = \sin \dfrac{\pi}{2} \cos 2x - \cos \dfrac{\pi}{2} \sin 2x$
$$= (1) \cos 2x - (0) \sin 2x$$
$$= \cos 2x - 0$$
$$= \cos 2x$$

(c) $\cos\left(\dfrac{\pi}{2} - 2x\right) = \cos\dfrac{\pi}{2}\cos 2x + \sin\dfrac{\pi}{2}\sin 2x$

$$= (0)\cos 2x + (1)\sin 2x$$
$$= \sin 2x$$

(d) $\cos 5\cos 10 - \sin 5\sin 10 = \cos(5 + 10) = \cos 15$

(e) $\cos(x + y) + \cos(x - y) = \cos x\cos y - \sin x\sin y + \cos x\cos y + \sin x\sin y$

$$= 2\cos x\cos y$$

3 (a) *Proof*: $\sin(\pi - A) = \sin\pi\cos A - \cos\pi\sin A$

$$= 0\cdot\cos A - (-1)\sin A$$
$$= \sin A$$

(b) *Proof*: $\cos(\pi - A) = \cos\pi\cos A + \sin\pi\sin A$

$$= (-1)\cos A + 0\cdot\sin A$$
$$= -\cos A$$

(c) *Proof*: $\tan\left(\dfrac{\pi}{2} - A\right) = \dfrac{\sin\left(\dfrac{\pi}{2} - A\right)}{\cos\left(\dfrac{\pi}{2} - A\right)} = \dfrac{\cos A}{\sin A} = \cot A$

(d) *Proof*: $\sec\left(\dfrac{\pi}{2} - t\right) = \dfrac{1}{\cos\left(\dfrac{\pi}{2} - t\right)} = \dfrac{1}{\sin t} = \csc t$

(e) *Proof*: $\dfrac{\tan(x + y) - \tan y}{1 + \tan(x + y)\tan y} = \tan[(x + y) - y] = \tan x$

(f) *Proof*: $\dfrac{\tan(x + y)}{\cot(x - y)} = \dfrac{\dfrac{\tan x + \tan y}{1 - \tan x\tan y}}{\dfrac{1 + \tan x\tan y}{\tan x - \tan y}}$

$$= \dfrac{\tan x + \tan y}{1 - \tan x\tan y}\cdot\dfrac{\tan x - \tan y}{1 + \tan x\tan y}$$

$$= \dfrac{\tan^2 x - \tan^2 y}{1 - \tan^2 x\tan^2 y}$$

(g) *Proof*: $\tan(225° - x) = \dfrac{\tan 225° - \tan x}{1 + \tan 225°\tan x}$

$$= \dfrac{1 - \tan x}{1 + (1)\tan x}$$

$$= \dfrac{1 - \tan x}{1 + \tan x}$$

(h) *Proof*: $\cot (x + y) = \dfrac{1}{\tan (x + y)} = \dfrac{1}{\dfrac{\tan x + \tan y}{1 - \tan x \tan y}} = \dfrac{1 - \tan x \tan y}{\tan x + \tan y}$

$$= \dfrac{1 - \dfrac{1}{\cot x} \cdot \dfrac{1}{\cot y}}{\dfrac{1}{\cot x} + \dfrac{1}{\cot y}} = \dfrac{\dfrac{\cot x \cot y - 1}{\cot x \cot y}}{\dfrac{\cot y + \cot x}{\cot x \cot y}} = \dfrac{\cot x \cot y - 1}{\cot x + \cot y}$$

(i) *Proof*: $\sin (A + B) - \sin (A - B) = (\sin A \cos B + \cos A \sin B) -$
$$(\sin A \cos B - \cos A \sin B)$$
$$= 2 \cos A \sin B$$

4 (a) $\tan (A - B) = -\frac{220}{21}$
(c) $\sin (k + t) = -\frac{24}{25}$
(e) $\tan (k - t) = -\frac{24}{7}$

(b) $\cos 75° = \frac{1}{2}\sqrt{2 - \sqrt{3}}$
(d) $\cos (k + t) = \frac{7}{25}$

Exercise 9-5.1 **1** (a) $\sin 2A = -\frac{120}{169}$; $\cos 2A = \frac{119}{169}$; $\tan 2A = -\frac{120}{119}$
(b) $\sin 2A = \frac{120}{169}$; $\cos 2A = -\frac{119}{169}$; $\tan 2A = -\frac{120}{119}$
(c) $\sin 2A = \frac{240}{289}$; $\cos 2A = -\frac{161}{289}$; $\tan 2A = -\frac{240}{161}$
(d) $\sin 2A = \dfrac{2r}{r^2 + 1}$; $\cos 2A = \dfrac{r^2 - 1}{r^2 + 1}$; $\tan 2A = \dfrac{2r}{r^2 - 1}$

2 $\tan 120° = -\sqrt{3}$

3 (a) *Proof*: $\sin 2x = 2 \sin x \cos x$
$$= \dfrac{2 \cos x}{\dfrac{1}{\sin x}} = \dfrac{2 \cos x}{\dfrac{1}{\sin x}} \cdot \dfrac{\cos x}{\cos x} = \dfrac{2 \cos^2 x}{\cot x}$$

(b) *Proof*: $\tan x \sin 2x = \tan x \cdot 2 \sin x \cos x$
$$= \dfrac{\sin x}{\cos x} \cdot 2 \sin x \cos x = 2 \sin^2 x$$

(c) *Proof*: $2 \cot 2A = \dfrac{2 \cos 2A}{\sin 2A}$
$$= \dfrac{2(\cos^2 A - \sin^2 A)}{2 \sin A \cos A}$$
$$= \dfrac{\cos A \cos A}{\sin A \cos A} - \dfrac{\sin A \sin A}{\sin A \cos A} = \cot A - \tan A$$

(d) *Proof*: $\dfrac{\cos 2A}{\cos A - \sin A} = \dfrac{\cos^2 A - \sin^2 A}{\cos A - \sin A}$
$$= \dfrac{(\cos A - \sin A)(\cos A + \sin A)}{\cos A - \sin A}$$
$$= \cos A + \sin A$$

(e) *Proof:* $\cos 10A = \cos (2 \cdot 5A)$
$$= 1 - 2 \sin^2 5A$$

(f) *Proof:* $\cos 2t = (\cos^2 t - \sin^2 t) \cdot 1$
$$= (\cos^2 t - \sin^2 t)(\cos^2 t + \sin^2 t)$$
$$= \cos^4 t - \sin^4 t$$

(g) *Proof:* $\cos^3 t - \sin^3 t = (\cos t - \sin t)(\cos^2 t + \cos t \sin t + \sin^2 t)$
$$= (\cos t - \sin t)(\cos^2 t + \sin^2 t + \cos t \sin t)$$
$$= (\cos t - \sin t)(1 + \tfrac{1}{2} \sin 2t)$$

(h) *Proof:* $\dfrac{1 - \cos 2x}{2} = \dfrac{1 - (1 - 2 \sin^2 x)}{2}$
$$= \dfrac{1 - 1 + 2 \sin^2 x}{2}$$
$$= \dfrac{2 \sin^2 x}{2}$$
$$= \sin^2 x$$

(i) *Proof:* $\cos^6 B - \sin^6 B = (\cos^2 B)^3 - (\sin^2 B)^3$
$$= (\cos^2 B - \sin^2 B)(\cos^4 B + \sin^2 B \cos^2 B + \sin^4 B)$$
$$= \cos 2B (\cos^4 B + 2 \sin^2 B \cos^2 B + \sin^4 B - \sin^2 B \cos^2 B$$
$$= \cos 2B [(\cos^2 B + \sin^2 B)^2 - \sin^2 B \cos^2 B]$$
$$= \cos 2B (1 - \sin^2 B \cos^2 B)$$

(j) *Proof:* $\cos^4 x - \sin^4 x = (\cos^2 x + \sin^2 x)(\cos^2 x - \sin^2 x)$
$$= 1(\cos^2 x - \sin^2 x)$$
$$= \cos 2x$$
$$= 2 \cos^2 x - 1$$

Exercise 9-6.1 **1** (a) $\sin \tfrac{1}{2}A = \tfrac{1}{5}\sqrt{5}$; $\cos \tfrac{1}{2}A = \tfrac{2}{5}\sqrt{5}$; $\tan \tfrac{1}{2}A = \tfrac{1}{2}$
(b) $\sin \tfrac{1}{2}A = \tfrac{5}{34}\sqrt{34}$; $\cos \tfrac{1}{2}A = -\tfrac{3}{34}\sqrt{34}$; $\tan \tfrac{1}{2}A = -\tfrac{5}{3}$
(c) $\sin \tfrac{1}{2}A = \tfrac{5}{34}\sqrt{34}$; $\cos \tfrac{1}{2}A = \tfrac{3}{34}\sqrt{34}$; $\tan \tfrac{1}{2}A = \tfrac{5}{3}$

2 (a) $\cos 15° = \tfrac{1}{2}\sqrt{2 + \sqrt{3}}$
(b) $\tan 22.5° = \sqrt{2} - 1$
(c) $\cot 75° = 2 - \sqrt{3}$
(d) $\sin \dfrac{\pi}{8} = \dfrac{\sqrt{2 - \sqrt{2}}}{2}$
(e) $\sec 105° = -2\sqrt{2 + \sqrt{3}}$
(f) $\sin 112.5° = \dfrac{\sqrt{2 + \sqrt{2}}}{2}$

(g) $\cos 157.5° = \dfrac{-\sqrt{2 + \sqrt{2}}}{2}$

(h) $\cos \dfrac{5\pi}{12} = \dfrac{\sqrt{2 - \sqrt{3}}}{2}$

3 (a) *Proof*:

$$(1 + \cos 2x)\left(1 + \tan \frac{x}{2} \tan x\right)^2 = (1 + \cos 2x)\left(1 + \frac{1 - \cos x}{\sin x} \cdot \frac{\sin x}{\cos x}\right)^2$$

$$= (1 + \cos 2x)\left(1 + \frac{1 - \cos x}{\cos x}\right)^2$$

$$= (1 + \cos 2x)\left(\frac{1}{\cos^2 x}\right)$$

$$= \frac{1 + 2\cos^2 x - 1}{\cos^2 x}$$

$$= 2$$

(b) *Proof*: $\sin \dfrac{x}{2} \cos \dfrac{x}{2} = \dfrac{\pm\sqrt{1 - \cos x}}{2} \cdot \dfrac{\pm\sqrt{1 + \cos x}}{2}$

$$= \sqrt{\frac{1^2 - \cos^2 x}{4}}$$

$$= \sqrt{\frac{\sin^2 x}{4}}$$

$$= \tfrac{1}{2} \sin x$$

(c) *Proof*: $\sec^2 \dfrac{x}{2} = \dfrac{1}{\cos^2 \dfrac{x}{2}}$

$$= \frac{1}{\left(\sqrt{\dfrac{1 + \cos x}{2}}\right)^2}$$

$$= \frac{2}{1 + \cos x}$$

$$= \frac{2 \cdot \dfrac{1}{\cos x}}{\dfrac{1 + \cos x}{\cos x}}$$

$$= \frac{2 \sec x}{1 + \sec x}$$

(d) *Proof*: $\left(1 - 2\sin\dfrac{A}{2}\right)\left(1 + \sin\dfrac{A}{2}\right)$

$$= \left(1 - 2\sqrt{\dfrac{1 - \cos A}{2}}\right)\left(1 + \sqrt{\dfrac{1 - \cos A}{2}}\right)$$

$$= 1 - \sqrt{\dfrac{1 - \cos A}{2}} - 2\left(\dfrac{1 - \cos A}{2}\right)$$

$$= 1 - \sqrt{\dfrac{1 - \cos A}{2}} - 1 + \cos A$$

$$= \cos A - \sqrt{\dfrac{1 - \cos A}{2}} = \cos A - \sin\dfrac{A}{2}$$

(e) *Proof*: $\cot\dfrac{A}{2} - \tan\dfrac{A}{2} = \dfrac{1}{\tan\dfrac{A}{2}} - \tan\dfrac{A}{2}$

$$= \dfrac{1 + \cos A}{\sin A} - \dfrac{1 - \cos A}{\sin A} = \dfrac{2\cos A}{\sin A}$$

$$= 2\cot A$$

(f) *Proof*: $2\cos^2\dfrac{A}{2} = 2\left(\pm\sqrt{\dfrac{1 + \cos A}{2}}\right)^2$

$$= 2 \cdot \dfrac{1 + \cos A}{2}$$

$$= 1 + \cos A$$

(g) *Proof*: $\sec\dfrac{x}{2} + \csc\dfrac{x}{2} = \dfrac{1}{\cos\dfrac{x}{2}} + \dfrac{1}{\sin\dfrac{x}{2}}$

$$= \dfrac{\sin\dfrac{x}{2} + \cos\dfrac{x}{2}}{\cos\dfrac{x}{2}\,\sin\dfrac{x}{2}}$$

$$= \dfrac{\sin\dfrac{x}{2} + \cos\dfrac{x}{2}}{\dfrac{(1 - \cos x)(1 + \cos x)}{2}}$$

$$= 2 \cdot \dfrac{\sin\dfrac{x}{2} + \cos\dfrac{x}{2}}{1 - \cos^2 x}$$

$$= \dfrac{2}{\sin^2 x}\left(\sin\dfrac{x}{2} + \cos\dfrac{x}{2}\right)$$

$$= 2\csc^2 x\left(\sin\dfrac{x}{2} + \cos\dfrac{x}{2}\right)$$

(h) *Proof*: $\cot \dfrac{B}{2} = \dfrac{1}{\tan \dfrac{B}{2}}$

$= \dfrac{1}{\dfrac{1 - \cos B}{\sin B}} = \dfrac{\sin B}{1 - \cos B}$

(i) *Proof*: $\cot^2 \dfrac{B}{2} = \dfrac{1}{\tan^2 \dfrac{B}{2}}$

$= \dfrac{1}{\left(\sqrt{\dfrac{1 - \cos B}{1 + \cos B}}\right)^2}$

$= \dfrac{1 + \cos B}{1 - \cos B}$

(j) *Proof*: $8 \sin^2 \tfrac{1}{2}A \cos^2 \tfrac{1}{2}A = 8 \cdot \dfrac{1 - \cos A}{2} \cdot \dfrac{1 + \cos A}{2}$

$= 8\left(\dfrac{1 - \cos^2 A}{4}\right)$

$= 2(1 - \cos^2 A)$

$= 2 \sin^2 A$

Review Exercise **1** (a) $\sin \theta$

(b) $\sin A \cos B + \cos A \sin B$

(c) $\dfrac{\tan A + \tan B}{1 - \tan A \tan B}$

(d) $2 \cos^2 A - 1$

(e) $\pm \sqrt{\dfrac{1 - \cos x}{2}}$

(f) $\dfrac{\sin C}{1 + \cos C}$

2 (a) $\sin 16°23'$

(b) $\cot 68°47'$

(c) $\cos (-110°)$

(d) $\csc (-20°)$

(e) $\sec 190°$

(f) $\tan 140°$

(g) $\cos 110°39'$

(h) $\sin (-410°)$

3 (a) *Proof*: $\cot x \sin x = \dfrac{\cos x}{\sin x} \cdot \sin x$

$= \cos x$

(b) *Proof*: $\sec A \tan A = \dfrac{1}{\cos A} \cdot \dfrac{\sin A}{\cos A}$

$= \dfrac{\sin A}{\cos^2 A}$

(c) *Proof*: $\tan A \sin A + \cos A = \dfrac{\sin A}{\cos A} \cdot \sin A + \cos A$

$$= \dfrac{\sin^2 A + \cos^2 A}{\cos A}$$

$$= \dfrac{1}{\cos A}$$

$$= \sec A$$

(d) *Proof*: $\tan A + \dfrac{\cos A}{1 + \sin A} = \dfrac{\sin A}{\cos A} + \dfrac{\cos A}{1 + \sin A}$

$$= \dfrac{\sin A + \sin^2 A + \cos^2 A}{\cos A(1 + \sin A)}$$

$$= \dfrac{\sin A + 1}{\cos A\,(1 + \sin A)}$$

$$= \dfrac{1}{\cos A}$$

$$= \sec A$$

(e) *Proof*: $\tan(45° + x) = \dfrac{\tan 45° + \tan x}{1 - \tan 45° \tan x}$

$$= \dfrac{1 + \tan x}{1 - \tan x}$$

(f) *Proof*: $\dfrac{1 + \cos 2C}{1 - \sin C} = \dfrac{1 + 1 - 2\sin^2 C}{1 - \sin C}$

$$= \dfrac{2(1 - \sin^2 C)}{1 - \sin C}$$

$$= \dfrac{2(1 + \sin C)(1 - \sin C)}{1 - \sin C}$$

$$= 2(1 + \sin C)$$

(g) *Proof*: $\dfrac{\sec^2 \frac{1}{2}x}{\tan^2 \frac{1}{2}x} = \dfrac{\dfrac{1}{\cos^2 \frac{1}{2}x}}{\left(\dfrac{\sin x}{1 + \cos x}\right)^2}$

$$= \dfrac{2}{1 + \cos x} \cdot \dfrac{(1 + \cos x)^2}{\sin^2 x}$$

$$= \dfrac{2}{\sin^2 x}(1 + \cos x)$$

$$= 2\csc^2 x(1 + \cos x)$$

(h)   *Proof:* $\dfrac{2\cos^2 x}{\sin^2 2x} = \dfrac{2\cos^2 x}{(2\sin x\cos x)^2}$

$$= \dfrac{\cos^2 x}{2\sin^2 x\cos^2 x}$$

$$= \dfrac{1}{2\sin^2 x}$$

$$= \tfrac{1}{2}\csc^2 x$$

(i)   *Proof:* $\dfrac{4\cos^4 \frac{1}{2}x}{\tan^2 x} = \dfrac{4\left(\pm\sqrt{\dfrac{1+\cos x}{2}}\right)^4}{\dfrac{\sin^2 x}{\cos^2 x}}$

$$= \dfrac{(1+\cos x)^2\cos^2 x}{\sin^2 x}$$

$$= \dfrac{(1+\cos x)(1+\cos x)\cos^2 x}{1-\cos^2 x}$$

$$= \dfrac{(1+\cos x)(1+\cos x)\cos^2 x}{(1+\cos x)(1-\cos x)}$$

$$= \dfrac{\cos^2 x(1+\cos x)}{1-\cos x}$$

$$= \dfrac{\cos^2 x}{\dfrac{1-\cos x}{1+\cos x}}$$

$$= \dfrac{\cos^2 x}{\left(\pm\sqrt{\dfrac{1-\cos x}{1+\cos x}}\right)^2}$$

$$= \dfrac{\cos^2 x}{\tan^2 \frac{1}{2}x}$$

4   (a) $\frac{1}{2}\sqrt{2-\sqrt{2}}$ (b) $-\frac{1}{2}\sqrt{2-\sqrt{2}}$

      (c) $\frac{204}{325}$ (d) $-\frac{21}{220}$

5   (a) $\sin k$ (b) $\tan 8M$

      (c) $\sin 5°$ (d) $\tan \left(\frac{35}{2}\right)°$ or $\tan 17.5°$

      (e) $\sin 90°$ (f) 0

Product and Sum Formulas

Objectives

Upon completion of Chapter 10, you will be able to:

1 **Write from memory the following identities (10-1):**

$\sin (A + B) + \sin (A - B) = 2 \sin A \cos B$

$\sin (A + B) - \sin (A - B) = 2 \cos A \sin B$

$\cos (A + B) + \cos (A - B) = 2 \cos A \cos B$

$\cos (A + B) - \cos (A - B) = -2 \sin A \sin B$

$\sin A + \sin B = 2 \sin \dfrac{A + B}{2} \cos \dfrac{A - B}{2}$

$\sin A - \sin B = 2 \cos \dfrac{A + B}{2} \sin \dfrac{A - B}{2}$

$\cos A + \cos B = 2 \cos \dfrac{A + B}{2} \cos \dfrac{A - B}{2}$

$\cos A - \cos B = -2 \sin \dfrac{A + B}{2} \sin \dfrac{A - B}{2}$

2 **Use the formulas in objective 1 to express products of sines or cosines as sums or differences of sines or cosines.** (10-1)

3 **Use the formulas in objective 1 to express sums or differences of sines or cosines as products of sines or cosines.** (10-1)

4 **Prove identities using the formulas in objective 1.** (10-1)

10/Product and Sum Formulas

10-1 Sums and Differences of Two Sines or Cosines
Although the formulas given in this chapter are not used as frequently as those given in Chapters 8 and 9, they can be studied on an optional basis. These formulas are used occasionally in a study of the calculus.

The first four identities, when read from left to right, allow us to express the product of a sine or cosine of one angle by the sine or cosine of another as one half of a sum or difference of sines or cosines.

10-1.1 $\qquad 2 \sin A \cos B = \sin (A + B) + \sin (A - B)$

10-1.2 $\qquad 2 \cos A \sin B = \sin (A + B) - \sin (A - B)$

10-1.3 $\qquad 2 \cos A \cos B = \cos (A + B) + \cos (A - B)$

10-1.4 $\qquad -2 \sin A \sin B = \cos (A + B) - \cos (A - B)$

The proof of each is left as an exercise to the student.

Example 10-1.1 Prove that $\sin (x + y) - \sin (x - y) = 2 \cos x \sin y$ (formula 10-1.2).

Proof $\qquad \sin (x + y) = \sin x \cos y + \cos x \sin y$

and

$$\sin (x - y) = \sin x \cos y - \cos x \sin y$$

Hence,

$$\sin (x + y) - \sin (x - y) = \sin x \cos y + \cos x \sin y$$
$$- (\sin x \cos y - \cos x \sin y)$$
$$= 2 \cos x \sin y$$

Example 10-1.2 Write $\cos A \sin 3A$ as a sum of sines.

Solution From formula 10-1.2 we have:

$$\cos A \sin 3A = \tfrac{1}{2} [\sin (A + 3A) - \sin (A - 3A)]$$
$$= \tfrac{1}{2} [\sin 4A - \sin (-2A)]$$
$$= \tfrac{1}{2} (\sin 4A + \sin 2A)$$

Example 10-1.3 Prove the identity $\sin 3A = \sin A + 2 \cos 2A \sin A$.

Proof $\qquad \sin A + 2 \cos 2A \sin A = \sin A + \sin (2A + A)$
$$- \sin (2A - A) \qquad \text{(from formula 10-1.2)}$$

$$= \sin A + \sin 3A - \sin A$$

$$= \sin 3A$$

In formulas 10-1.1 through 10-1.4, if we let $A + B = C$ and $A - B = D$, then $C + D = 2A$ and $C - D = 2B$. Hence, we get the following four identities:

10-1.5 $\qquad \sin C + \sin D = 2 \sin \dfrac{C + D}{2} \cos \dfrac{C - D}{2}$

10-1.6 $\qquad \sin C - \sin D = 2 \cos \dfrac{C + D}{2} \sin \dfrac{C - D}{2}$

10-1.7 $\qquad \cos C + \cos D = 2 \cos \dfrac{C + D}{2} \cos \dfrac{C - D}{2}$

10-1.8 $\qquad \cos C - \cos D = -2 \sin \dfrac{C + D}{2} \sin \dfrac{C - D}{2}$

Example 10-1.4 Express $\cos 40° - \cos 60°$ as a product.

Solution $\qquad \cos 40° - \cos 60° = -2 \sin \dfrac{40° + 60°}{2} \sin \dfrac{40° - 60°}{2}$

$$= -2 \sin 50° \sin (-10°)$$

$$= 2 \sin 50° \sin 10°$$

Example 10-1.5 Prove the identity

$$\frac{\sin A + \sin B}{\cos A + \cos B} = \tan \frac{A + B}{2}$$

Proof $\qquad \dfrac{\sin A + \sin B}{\cos A + \cos B} = \dfrac{2 \sin \dfrac{A + B}{2} \cos \dfrac{A - B}{2}}{2 \cos \dfrac{A + B}{2} \cos \dfrac{A - B}{2}}$

$$= \frac{\sin \dfrac{A + B}{2}}{\cos \dfrac{A + B}{2}}$$

$$= \tan \frac{A + B}{2}$$

Exercise 10-1.1 **1** Express each product as a sum.

(a) $\cos 4x \cos x$

(b) $\sin A \cos 2A$

(c) $2 \cos 3A \sin 4A$

(d) $2 \sin \dfrac{5\pi}{6} \cos \dfrac{2\pi}{3}$

2 Express each sum as a product.

(a) $\sin \dfrac{5\pi}{6} + \sin \dfrac{2\pi}{3}$

(b) $\cos 3A + \cos 5A$

(c) $\cos 2A - \cos 3A$

(d) $\sin 70° - \sin 10°$

3 Prove the following identities:

(a) $\sin A - \cos 2A - \sin 3A = -\cos 2A \,(2 \sin A + 1)$

(b) $\dfrac{\cos A + \cos B}{\sin A - \sin B} = \cot \dfrac{A - B}{2}$

(c) $\sin (A + 45°) + \sin (A - 45°) = \sqrt{2} \sin A$

(d) $\dfrac{\sin 10x - \sin 4x}{\sin 4x + \sin 2x} = \dfrac{\cos 7x}{\cos x}$

(e) $\dfrac{\sin 7A + \sin A + \sin 4A}{\cos 7A + \cos A + \cos 4A} = \tan 4A$

(f) $\sin 4x + \sin 2x = 2 \sin 3x \cos x$

(g) $\dfrac{\cos x + \cos y}{\sin x - \sin y} = \cot \tfrac{1}{2}(x - y)$

(h) $\dfrac{\cos A - \cos B}{\sin A - \sin B} = -\tan \frac{1}{2}(A + B)$

(i) $\dfrac{\sin 10A + \sin 6A}{\sin 12A + \sin 4A} - \dfrac{\sin 5A - \sin A}{\sin 7A + \sin A} = \dfrac{2 \sin 2A}{\sin 8A}$

Review Exercise   **1**   Fill in the blanks by using the product and sum formulas.

(a) $\cos (C + D) + \cos (C - D) = 2$ _____

(b) $\cos A + \cos B = 2 \cos$ _____ $\cos$ _____

(c) $2 \cos \dfrac{A + B}{2} \sin \dfrac{A - B}{2} =$ _____

(d) $2 \cos A \sin B = \sin$ _____ $- \sin$ _____

2   Express each product as a sum or difference of sines or cosines.

(a) $\sin \frac{3}{4}x \cos \frac{1}{4}x$ 　　　　　　　　　(b) $2 \sin 4A \cos 2A$

(c) $3 \cos \sqrt{3}A \cos 2\sqrt{3}A$ 　　　　　(d) $\sin 2A \sin 3A$

3   Express each sum or difference as a product of sines or cosines.

(a) $\sin 7x - \sin 3x$ 　　　　　　　　　(b) $\cos 38° - \cos 10°$

(c) $\sin \frac{7}{2}A + \sin \frac{3}{2}A$ 　　　　　　(d) $\cos 4\sqrt{5}B + \cos 2\sqrt{5}B$

4   Prove each of the following identities.

(a)   Prove each of the identities given in formulas 10-1.1 through 10-1.8.

(b) $\dfrac{\cos A - \cos B}{\cos A + \cos B} = -\tan \frac{1}{2}(A + B) \tan \frac{1}{2}(A - B)$

(c) $\sin^2 3A - \sin^2 A = 2 \sin^2 2A \cos 2A$

(d) $\dfrac{\cos x + \cos 2x + \cos 3x}{\sin x + \sin 2x + \sin 3x} = \cot 2x$

(e) $\cos (60° - A) + \cos (60° + A) = \cos A$

Answers

Exercise 10-1.1

1 (a) $\cos 4x \cos x = \frac{1}{2}(\cos 5x + \cos 3x)$
(b) $\sin A \cos 2A = \frac{1}{2}(\sin 3A - \sin A)$
(c) $2 \cos 3A \sin 4A = \sin 7A + \sin A$
(d) $2 \sin \dfrac{5\pi}{6} \cos \dfrac{2\pi}{3} = -\frac{1}{2}$

2 (a) $2 \sin \dfrac{3\pi}{4} \cos \dfrac{\pi}{12}$ (b) $2 \cos 4A \cos A$

(c) $2 \sin \dfrac{5A}{2} \sin \dfrac{A}{2}$ (d) $2 \cos 40° \sin 30°$

3 (a) *Proof*: $-\cos 2A (2 \sin A + 1) = -2 \cos 2A \sin A - \cos 2A$
$= -(\sin 3A - \sin A) - \cos 2A$
$= \sin A - \cos 2A - \sin 3A$

(b) *Proof*: $\dfrac{\cos A + \cos B}{\sin A - \sin B} = \dfrac{2 \cos \dfrac{A + B}{2} \cos \dfrac{A - B}{2}}{2 \cos \dfrac{A + B}{2} \sin \dfrac{A - B}{2}}$

$= \cot \dfrac{A - B}{2}$

(c) *Proof*:

$\sin (A + 45°) + \sin (A - 45°) = 2 \sin \dfrac{A + 45° + A - 45°}{2}$

$\cos \dfrac{A + 45° - A + 45°}{2}$

$= 2 \sin A \cos 45°$
$= \sqrt{2} \sin A$

(d) *Proof*: $\dfrac{\sin 10x - \sin 4x}{\sin 4x + \sin 2x} = \dfrac{2 \cos \dfrac{10x + 4x}{2} \sin \dfrac{10x - 4x}{2}}{2 \sin \dfrac{4x + 2x}{2} \cos \dfrac{4x - 2x}{2}}$

$$= \frac{\cos 7x \sin 3x}{\sin 3x \cos x}$$

$$= \frac{\cos 7x}{\cos x}$$

(e) *Proof*:

$$\frac{\sin 7A + \sin A + \sin 4A}{\cos 7A + \cos A + \cos 4A} = \frac{2 \sin \dfrac{7A + A}{2} \cos \dfrac{7A - A}{2} + \sin 4A}{2 \cos \dfrac{7A + A}{2} \cos \dfrac{7A - A}{2} + \cos 4A}$$

$$= \frac{2 \sin 4A \cos 3A + \sin 4A}{2 \cos 4A \cos 3A + \cos 4A}$$

$$= \frac{\sin 4A \, (2 \cos 3A + 1)}{\cos 4A \, (2 \cos 3A + 1)}$$

$$= \frac{\sin 4A}{\cos 4A}$$

$$= \tan 4A$$

(f) *Proof*: $\sin 4x + \sin 2x = 2 \sin \dfrac{4x + 2x}{2} \cos \dfrac{4x - 2x}{2}$

$$= 2 \sin 3x \cos x$$

(g) *Proof*: $\dfrac{\cos x + \cos y}{\sin x - \sin y} = \dfrac{2 \cos \dfrac{x + y}{2} \cos \dfrac{x - y}{2}}{2 \cos \dfrac{x + y}{2} \sin \dfrac{x - y}{2}}$

$$= \cot \tfrac{1}{2}(x - y)$$

(h) *Proof*: $\dfrac{\cos A - \cos B}{\sin A - \sin B} = \dfrac{-2 \sin \dfrac{A + B}{2} \sin \dfrac{A - B}{2}}{2 \cos \dfrac{A + B}{2} \sin \dfrac{A - B}{2}}$

$$= -\tan \tfrac{1}{2}(A + B)$$

(i) *Proof*:

$$\frac{\sin 10A + \sin 6A}{\sin 12A + \sin 4A} - \frac{\sin 5A - \sin A}{\sin 7A + \sin A} = \frac{2 \sin 8A \cos 2A}{2 \sin 8A \cos 4A} - \frac{2 \cos 3A \sin 2A}{2 \sin 4A \cos 3A}$$

$$= \frac{2 \sin 4A \cos 2A - 2 \sin 2A \cos 4A}{2 \cos 4A \sin 4A}$$

$$= \frac{\sin 6A + \sin 2A + \sin 2A - \sin 6A}{\sin 8A}$$

$$= \frac{2 \sin 2A}{\sin 8A}$$

Review Exercise **1** (a) $\cos (C + D) + \cos (C - D) = 2 \cos C \cos D$

(b) $\cos A + \cos B = 2 \cos \dfrac{A + B}{2} \cos \dfrac{A - B}{2}$

(c) $2 \cos \dfrac{A + B}{2} \sin \dfrac{A - B}{2} = \sin A - \sin B$

(d) $2 \cos A \sin B = \sin (A + B) - \sin (A - B)$

2 (a) $\frac{1}{2}(\sin x + \sin \frac{1}{2}x)$ (b) $\sin 6A + \sin 2A$

(c) $\frac{3}{2}(\cos 3\sqrt{3}A + \cos \sqrt{3}A)$ (d) $-\frac{1}{2}(\cos 5A - \cos A)$

3 (a) $\sin 7x - \sin 3x = 2 \cos 5x \sin 2x$

(b) $\cos 38° - \cos 10° = -2 \sin 24° \sin 14°$

(c) $\sin \frac{7}{2}A + \sin \frac{3}{2}A = 2 \sin \frac{5}{2}A \cos A$

(d) $\cos 4\sqrt{5}B + \cos 2\sqrt{5}B = 2 \cos 3\sqrt{5}B \cos \sqrt{5}B$

4 (a) *Prove*: $2 \sin A \cos B = \sin (A + B) + \sin (A - B)$

Proof: $\sin (A + B) = \sin A \cos B + \cos A \sin B$
$\qquad\qquad \sin (A - B) = \sin A \cos B - \cos A \sin B$

Add these two equations to get
$\qquad \sin (A + B) + \sin (A - B) = 2 \sin A \cos B$

Prove: $2 \cos A \sin B = \sin (A + B) - \sin (A - B)$

Proof: See Example 10-1.1.

Prove: $2 \cos A \cos B = \cos (A + B) - \cos (A - B)$

Proof: $\cos (A + B) = \cos A \cos B - \sin A \sin B$
$\qquad\qquad \cos (A - B) = \cos A \cos B + \sin A \sin B$

Add these two equations to get
$\qquad \cos (A + B) + \cos (A - B) = 2 \cos A \cos B$

Prove: $-2 \sin A \sin B = \cos (A + B) - \cos (A - B)$

Proof: $\cos (A + B) = \cos A \cos B - \sin A \sin B$
$\qquad\qquad \cos (A - B) = \cos A \cos B + \sin A \sin B$
$\qquad\qquad \cos (A + B) - \cos (A - B) = -2 \sin A \sin B$

Prove: Formulas 10-1.5 through 10-1.8.

Proof: If $A + B = C$ and $A - B = D$, then $C + D = 2A$ and
$\qquad C - D = 2B.$
$\qquad\qquad \sin (A + B) + \sin (A - B) = \sin C + \sin D$
$\qquad\qquad \sin (A + B) + \sin (A - B) = 2 \sin A \cos B$
$$= 2 \sin \frac{C + D}{2} \cos \frac{C - D}{2}$$

Hence,

$$\sin C + \sin D = 2 \sin \frac{C + D}{2} \cos \frac{C - D}{2} \qquad (10\text{-}1.5)$$

$$\sin (A + B) - \sin (A - B) = \sin C - \sin D$$
$$\sin (A + B) - \sin (A - B) = 2 \cos A \sin B$$
$$= 2 \cos \frac{C + D}{2} \sin \frac{C - D}{2} \qquad (10\text{-}1.6)$$

$$\cos (A + B) + \cos (A - B) = \cos C + \cos D$$
$$\cos (A + B) + \cos (A - B) = 2 \cos A \cos B$$
$$= 2 \cos \frac{C + D}{2} \cos \frac{C - D}{2}$$
$$\cos C + \cos D = 2 \cos \frac{C + D}{2} \cos \frac{C - D}{2} \qquad (10\text{-}1.7)$$
$$\cos (A + B) - \cos (A - B) = \cos C - \cos D$$
$$\cos (A + B) - \cos (A - B) = -2 \sin A \sin B$$
$$= -2 \sin \frac{C + D}{2} \sin \frac{C - D}{2}$$
$$\cos C - \cos D = -2 \sin \frac{C + D}{2} \sin \frac{C - D}{2} \qquad (10\text{-}1.8)$$

(*b*) *Proof*:
$$\frac{\cos A - \cos B}{\cos A + \cos B} = \frac{-2 \sin \frac{1}{2}(A + B) \sin \frac{1}{2}(A - B)}{2 \cos \frac{1}{2}(A + B) \cos \frac{1}{2}(A - B)}$$
$$= -\tan \tfrac{1}{2}(A + B) \tan \tfrac{1}{2}(A - B)$$

(*c*) *Proof*: $\sin^2 3A - \sin^2 A = (\sin 3A + \sin A)(\sin 3A - \sin A)$
$$= \left(2 \sin \frac{3A + A}{2} \cos \frac{3A - A}{2} \right)$$
$$\left(2 \cos \frac{3A + A}{2} \sin \frac{3A - A}{2} \right)$$
$$= (2 \sin 2A \cos A)(2 \cos 2A \sin A)$$
$$= (2 \sin 2A \cos 2A)(2 \sin A \cos A)$$
$$= (2 \sin 2A \cos 2A)(\sin 2A)$$
$$= 2 \sin^2 2A \cos 2A$$

(*d*) *Proof*:
$$\frac{\cos x + \cos 2x + \cos 3x}{\sin x + \sin 2x + \sin 3x} = \frac{\cos 2x + 2 \cos \dfrac{4x}{2} \cos \dfrac{2x}{2}}{\sin 2x + 2 \sin \dfrac{4x}{2} \cos \dfrac{2x}{2}}$$
$$= \frac{\cos 2x \, (1 + 2 \cos x)}{\sin 2x \, (1 + 2 \cos x)}$$
$$= \cot 2x$$

(*e*) *Proof*: $\cos (60° - A) + \cos (60° + A) = 2 \cos 60° \cos A$
$$= 2 \cdot \tfrac{1}{2} \cdot \cos A$$
$$= \cos A$$

Trigonometric Equations

11

Objectives

Upon completion of Chapter 11, you will be able to:

1   Solve basic trigonometric equations of the type $T(\theta) = a$, where T is a trigonometric function, θ is any angle, and a is any real number. (11-2)

2   Solve trigonometric equations of the first degree involving one trigonometric function. (11-3)

3   Solve trigonometric equations of the second degree involving one trigonometric function. (11-4)

4   Solve trigonometric equations that involve more than one trigonometric function. (11-5)

5   Solve trigonometric equations that involve functions of a double angle or functions of a half angle. (11-6)

11/Trigonometric Equations

11-1 Introduction

In Chapter 9, it was pointed out that an *identity* is an equation that is true for all values of the variable for which both sides are defined. A *conditional equation* is an equation which is true for only a limited number of values of the variable. In this chapter we will consider methods for solving several types of conditional trigonometric equations.

To solve trigonometric equations, we use both algebra and trigonometry. Some points to remember are:

1 No single method works for all equations.
2 When you take the square root of both sides of an equation or divide both sides by an expression, be sure not to lose a solution to the original equation.
3 When squaring both sides of an equation or multiplying both sides by a variable, be sure to check your solutions in the original equation since some extraneous roots may have been introduced.
4 When θ is a solution to a trigonometric equation, then $\theta + 2n\pi$, $n \in Z$, is also a solution to the equation. In this chapter only solutions for θ, where $0° \leq \theta \leq 360°$, will be given in the solution set.

11-2 Basic Equations of the Form $T(\theta) = a$

The simplest type of trigonometric equation is the type which involves only one function and a real number. Examples of equations in this form are $\cos \theta = \frac{1}{2}$, $\sin \theta = \sqrt{\frac{3}{2}}$, and $\tan \theta = 0.5678$. If the solutions are special angles, quadrantal angles, or angles that have special angles as related angles, then the solutions can be written immediately upon inspection. Other cases are solved by giving approximate solutions from a table of values for the trigonometric functions.

The following examples illustrate types of basic equations and techniques for solving them. The solutions may be checked by quick inspection.

Example 11-2.1 Solve $\tan \theta = \sqrt{3}$.

Solution Since this is a value for one of the special angles, we know that $\theta = 60°, 240°$. Recall that we are giving solutions for θ where $0° \leq \theta \leq 360°$. Hence, our solution set (SS) may be written as

$$SS = \{60°, 240°\}$$

or

$$\text{SS} = \left\{ \frac{\pi}{3}, \frac{4\pi}{3} \right\} \qquad \text{where } 0 \le \theta \le 2\pi$$

Example 11-2.2 Solve $\cot t = 0$ for t.

Solution If $\cot t = 0$, SS = $\{90°, 270°\}$.

Example 11-2.3 Solve $\cos t = 0.8387$ for t.

Solution Using Table C-1, we see that if $t = 33°$ or $t = 327°$, then $\cos t = 0.8387$. Hence, SS = $\{33°, 327°\}$.

Example 11-2.4 Solve $\sin 2x = 0$ for x.

Solution $2x = 0°,\ 180°,\ 360°,\ 720°$

Hence, $x = 0°,\ 90°,\ 180°,\ 360°$, and

$\text{SS} = \{0°,\ 90°,\ 180°,\ 360°\}$

(Note that since $0° \le x \le 360°$, $0° \le 2x \le 720°$.)

Exercise 11-2.1 Solve the following equations. Use Table C-1 to find approximate solutions, to the nearest degree.

1 $2 \cos x = 1$

2 $2 \sin t + \sqrt{3} = 0$

3 $\tan A + \sqrt{3} = 0$

4 $\cot A - \dfrac{\sqrt{3}}{3} = 0$

5 $3 \tan B = 6$

6 $\cos x = 0$

7 $\tan x = 0$

8 $\tan A = -1$

9 $\sec x = -2$

10 $\csc A = \dfrac{2}{\sqrt{3}}$

11 $\sin A = 0.3420$

12 $\tan A = -2.748$

11-3 First-Degree Equations of One Function

To solve a first-degree trigonometric equation that contains only one function, perform the following steps:

STEP 1 Use algebraic techniques to get the trigonometric function on one side of the equation.

STEP 2 Use algebraic techniques to write everything else on the other side of the equation.

STEP 3 Use trigonometry to obtain solutions to the equation.

You should substitute your possible solutions into the original equation to check them.

Example 11-3.1 Solve $3 \sin x - 2 = \sin x$.

Solution
$$3 \sin x - 2 = \sin x$$
$$2 \sin x - 2 = 0$$
$$2 \sin x = 2$$
$$\sin x = 1$$
$$SS = \{90°\}$$

Check $3 \sin 90° - 2 = 3 \cdot 1 - 2 = 1$. The other side is $\sin 90° = 1$.

Example 11-3.2 Solve $2\sqrt{2} \cos x - 1 = -2 + \sqrt{2} \cos x$.

Solution
$$2\sqrt{2} \cos x - 1 = -2 + \sqrt{2} \cos x$$
$$2\sqrt{2} \cos x - \sqrt{2} \cos x = -2 + 1$$
$$\sqrt{2} \cos x = -1$$
$$\cos x = -\dfrac{1}{\sqrt{2}}$$
$$SS = \{135°, 225°\}$$

Check $2\sqrt{2} \cos 135° - 1 = 2\sqrt{2}\left(-\dfrac{1}{\sqrt{2}}\right) - 1 = -3$ while $-2 + \sqrt{2} \cos 135° =$ -3. Also $2\sqrt{2} \cos 225° - 1 = -3$ and $-2 + \sqrt{2} \cos 225° = -3$.

Exercise 11-3.1 Solve the following equations for the angles between 0° and 360°. Use Table C-1 for approximations to the nearest degree. All denominators are non-zero.

1 $5 \cos x + 2 = 7$

2 $3(\sec x - 1) = 5 - \sec x$

3 $\dfrac{1}{\tan x} + \dfrac{1}{2} = \dfrac{5}{6 \tan x} + \dfrac{1}{3}$

4 $\dfrac{1}{\sin x} + \dfrac{1}{\sin x - 1} = \dfrac{5}{\sin x - 1}$

5 $2 \csc x - 1 = 3$

6 $7(4 + 8 \cot x) - 3 = 6 \cot x + 75$

7 $\dfrac{15 \cos x + 12}{3} - 2 = 3 \cos x + 2$

8 $\dfrac{1}{\sec x} + \dfrac{2}{\sec x} = 3 - \dfrac{3}{\sec x}$

9 $\dfrac{8 \sin x + 10}{5} + \dfrac{6 \sin x + 1}{4} = 2 \sin x + 3$

10 $2 \cos^2 x - \sin x = 1 + 2 \cos^2 x$

11-4 Second-Degree Equations of One Function

Solving second-degree trigonometric equations involving only one function calls for the student to employ the same tactics used in solving algebraic quadratic equations. That is, use the same tools you had to solve

$$ax^2 + bx + c = 0 \qquad a \neq 0$$

to solve $aT^2(\theta) + bT(\theta) + c = 0$, $a \neq 0$ and T is a trigonometric function. Remember that if $aT^2(\theta) + bT(\theta) + c$ does not factor, use the quadratic formula

$$T(\theta) = \frac{-b \pm \sqrt{b^2 - 4ac}}{2a}$$

The examples below illustrate these methods. You should always check your answers to see that they are solutions.

Example 11-4.1 Solve $2 \sin^2 \theta = \frac{1}{2}$.

Solution

$$2 \sin^2 \theta = \frac{1}{2}$$

$$\sin^2 \theta = \frac{1}{4}$$

$$\sin \theta = \pm \frac{1}{2}$$

$$SS = \{30°, 150°, 210°, 330°\}$$

Example 11-4.2 Solve $1 - 2 \cos x + \cos^2 x = 1 + 3 \cos^2 x$.

Solution

$$1 - 2 \cos x + \cos^2 x = 1 + 3 \cos^2 x$$

$$-2 \cos x - 2 \cos^2 x = 0$$

$$2 \cos x + 2 \cos^2 x = 0$$

$$2 \cos x (1 + \cos x) = 0$$

$$2 \cos x = 0 \qquad \text{or} \qquad 1 + \cos x = 0$$

$$\cos x = 0 \qquad\qquad\qquad \cos x = -1$$

$$x = 90°, 270° \qquad\qquad\qquad x = 180°$$

$$SS = \{90°, 180°, 270°\}$$

Example 11-4.3 Solve $1 - 2 \sin^2 A - \sin A = 0$.

Solution
$$1 - 2 \sin^2 A - \sin A = 0$$
$$2 \sin^2 A + \sin A - 1 = 0$$
$$(2 \sin A - 1)(\sin A + 1) = 0$$

$2 \sin A - 1 = 0$ or $\sin A + 1 = 0$

$2 \sin A = 1$ $\sin A = -1$

$\sin A = \frac{1}{2}$ $A = 270°$

$A = 30°, 150°$

$SS = \{30°, 150°, 270°\}$

Example 11-4.4 Solve $3 \tan^2 A = 1 - 4 \tan A$.

Solution Since $3 \tan^2 A + 4 \tan A - 1 = 0$ won't factor, use the quadratic formula.

$$\tan A = \frac{-4 \pm \sqrt{4^2 - 4(3)(-1)}}{2(3)}$$

$$\tan A = \frac{-4 \pm \sqrt{28}}{6} = \frac{-4 \pm 5.292}{6}$$

$$\tan A = \frac{1.292}{6} \quad \text{or} \quad \tan A = \frac{-9.292}{6}$$

$$\tan A = 0.215 \quad \text{or} \quad \tan A = -1.549$$

$$SS = \{12°, 123°, 192°, 303°\}$$

Exercise 11-4.1 Solve the following equations. Use Table C-1 to give approximations to the nearest degree.

1 $\tan^2 x = 2$

2 $4 \sin^2 x = 3$

3 $3 \sec^2 A = 4$

4 $\cot^2 A = 1$

5 $\sin^2 A - 2 \sin A = 0$

6 $\tan^2 x - \tan x = 0$

7 $3 \cos^2 x - 4 \cos x + 1 = 0$

8 $\cot^2 A - \cot A - 3 = 0$

9 $3 \csc^2 x - 2 \csc x = 1$

10 $\cos^2 x - \cos x = 6$

11 $\tan^4 A + 3 \tan^2 A + 2 = 0$ **12** $2 \sin^2 A + 5 \sin A + 3 = 0$

13 $2 \csc^2 x - \csc x = 1$ **14** $3 \cot^2 A + 2 \cot A = 3$

15 $\sec^2 A - 4 = 0$ **16** $\tan^2 t - \tan t = 6$

17 $\tan^2 t = \sqrt{3}$ **18** $2 \cos^2 t = -1 + 3 \cos t$

11-5 Equations with More Than One Function

The solutions to trigonometric equations containing more than one function are found by using algebraic techniques. Many times, however, trigonometric identities must be substituted in order to get the equation in a solvable form. For example, the equation $3 \tan^2 A = \sec^2 A$ can be solved by changing either $\tan^2 A$ to $\sec^2 A - 1$ or changing $\sec^2 A$ to $\tan^2 A + 1$ in order to get an equation with only one function. Several examples follow which further exhibit some useful techniques in solving such equations.

Example 11-5.1 Solve $\sin x = 2 \cos x$.

Solution Square both sides to obtain

$$\sin^2 x = 4 \cos^2 x \qquad \text{(read again point 3 in Section 11-1)}$$

$$1 - \cos^2 x = 4 \cos^2 x$$

$$1 = 5 \cos^2 x$$

$$\tfrac{1}{5} = \cos^2 x$$

$$\frac{\pm\sqrt{5}}{5} = \cos x$$

$$\frac{\pm 2.236}{5} = \cos x$$

$$\pm 0.447 = \cos x$$

$$x = 63°, 117°, 243°, 297°$$

Checking these, we see that 117° and 297° *are extraneous solutions*. Hence, SS = {63°, 243°}.

Example 11-5.2 Solve $4 \sin t \cos t - 2 \cos t + 2 \sin t = 1$.

Solution
$$4 \sin t \cos t - 2 \cos t + 2 \sin t - 1 = 0$$
$$2 \cos t (2 \sin t - 1) + 2 \sin t - 1 = 0$$
$$(2 \sin t - 1)(2 \cos t + 1) = 0$$

$$2 \sin t - 1 = 0 \qquad\qquad 2 \cos t + 1 = 0$$
$$\sin t = \tfrac{1}{2} \qquad\qquad\qquad \cos t = -\tfrac{1}{2}$$
$$t = 30°,150° \qquad\qquad\quad t = 120°,240°$$

$$\text{SS} = \{30°,120°,150°,240°\}$$

Example 11-5.3 Solve $\tan x - 2 \sec x = 3$.

Solution
$$\tan x = 2 \sec x + 3$$
$$\tan^2 x = 4 \sec^2 x + 12 \sec x + 9$$
$$\sec^2 x - 1 = 4 \sec^2 x + 12 \sec x + 9$$
$$3 \sec^2 x + 12 \sec x + 10 = 0$$
$$\sec x = \frac{-12 \pm \sqrt{(12)^2 - 4(3)(10)}}{2(3)}$$
$$= \frac{-12 \pm \sqrt{24}}{6} = \frac{-12 \pm 2\sqrt{6}}{6}$$
$$= \frac{-6 \pm \sqrt{6}}{3} = \frac{-6 \pm 2.449}{3}$$

$$\sec x = \frac{-8.449}{3} \qquad\qquad \sec x = \frac{-3.551}{3}$$
$$\sec x = -2.816 \qquad\qquad \sec x = -1.184$$
$$x \approx 111°,249° \qquad\qquad x \approx 148°,212°$$

Since both sides of the equation were squared, we check the solutions. From Table C-1, $\tan 111° = -2.605$ and $2 \sec 111° = 2(-2.816) = -5.632$.

Hence, $\tan x - 2 \sec x = 3$, and $-2.605 + 5.632 \approx 3$. Thus $x \approx 111°$ is a solution.

$$\tan 249° = 2.605 \qquad 2 \sec 249° = 2(-2.816)$$
$$= -5.632$$

Since $2.605 - 5.632 \neq 3$, $x = 249°$ is not a solution. Similarly, $x = 148°$ is a solution while $x = 212°$ is not a solution. Hence, $\text{SS} = \{111°,148°\}$.

Exercise 11-5.1 Solve the following equations. Use Table C-1 for approximations to the nearest degree.

1 $2 \sin A \cos A + \cos A = 0$

2 $4 \sin x \cos^2 x - \sin x = 0$

3 $\tan x \sin x - \sqrt{3} \sin x = 0$

4 $\dfrac{1 - \sin x}{\cos x} = \cos x$

5 $\sin x = -\cos x$

6 $2 \tan x - 1 = \sec^2 x$

7 $3 \sin^2 A = 2 \cos A + 2$

8 $\csc^2 B = \cot^2 B + 1$

9 $\cos^3 t \csc^3 t = 4 \cot^2 t$

10 $\sec A + 2 \cos A = 3$

11 $\sec^2 x = \tan^2 x + 2$

12 $\cot x = \dfrac{\sin x}{1 + \cos x}$

One type of equation which involves two functions is the equation of the form

$a \sin x + b \cos x = c$

Equations of this type may be solved by using the identity

$\sin^2 x + \cos^2 x = 1$

The following example illustrates how to solve such equations.

Example 11-5.4 Solve $\sin x + \sqrt{3} \cos x = \sqrt{2}$.

Solution

$$\sin x = \sqrt{2} - \sqrt{3} \cos x$$

$$\sin^2 x = 2 - 2\sqrt{6} \cos x + 3 \cos^2 x$$

$$1 - \cos^2 x = 2 - 2\sqrt{6} \cos x + 3 \cos^2 x$$

$$0 = 1 - 2\sqrt{6} \cos x + 4 \cos^2 x$$

$$\cos x = \frac{2\sqrt{6} \pm \sqrt{24 - 4(4)1}}{8}$$

$$\cos x = \frac{2\sqrt{6} \pm \sqrt{8}}{8} = \frac{2(2.449) \pm 2.828}{8}$$

$$\cos x = \frac{4.898 + 2.828}{8} \qquad \cos x = \frac{4.898 - 2.828}{8}$$

$$\cos x \approx 0.966 \qquad\qquad \cos x \approx 0.259$$

$$x = 15°, 345° \qquad\qquad x = 75°, 285°$$

The student should verify by substitution that 15° and 285° are not solutions while 75° and 345° are solutions. Hence, SS = {75°,345°}.

Exercise 11-5.2 Solve the following equations.

1 $3 \sin A - \cos A = 4$

2 $\sin A + \cos A = 0$

3 $\sin x - \cos x = 1$

4 $\sqrt{2} \sin x - \cos x = -1$

5 $2 \cos x - \sin x = 1$

6 $\sqrt{3} \sin x + \cos x = \sqrt{2}$

11-6 Equations with Double Angles or Half Angles

Quite frequently you may encounter equations which involve double angles or half angles. In order to solve such equations, you should use the identities for double angles and half angles. Note, however, that at the beginning of this chapter we agreed to list only those solutions between 0° and 360°. Hence, if $0° \leq x \leq 360°$, then the solutions for $\frac{1}{2}x$ will be given by $0° \leq \frac{1}{2}x \leq 180°$, and solutions for $2x$ will be given by $0° \leq 2x \leq 720°$.

Several examples follow which indicate how equations containing double angles or half angles can be solved. Some solutions are checked while others are left for the student to verify.

Example 11-6.1 Solve $\sin^2 \frac{1}{2}x = \cos^2 x$.

Solution

$$\sin^2 \tfrac{1}{2}x = \cos^2 x$$

$$\left(\pm \sqrt{\frac{1 - \cos x}{2}} \right)^2 = \cos^2 x$$

$$\frac{1 - \cos x}{2} = \cos^2 x$$

$$1 - \cos x = 2 \cos^2 x$$

$$2 \cos^2 x + \cos x - 1 = 0$$

$$(2 \cos x - 1)(\cos x + 1) = 0$$

$$2 \cos x - 1 = 0 \qquad\qquad \cos x + 1 = 0$$

$$\cos x = \tfrac{1}{2} \qquad\qquad \cos x = -1$$

$$x = 60°, 300° \qquad\qquad x = 180°$$

$$SS = \{60°, 180°, 300°\}$$

Example 11-6.2 Solve $\sin 4A = 1$.

Solution Since $0° \leq A \leq 360°$, $0 \leq 4A \leq 1440°$. If $\sin 4A = 1$, $4A = 90°, 450°, 810°,$ 1170°.

$$SS = \{22.5°,112.5°,202.5°,292.5°\}$$

Check for 202.5°: $\sin 4(202.5°) = \sin 810° = \sin 90° = 1$.

Example 11-6.3 Solve $\cos 2x - \cos^2 x = 0$.

Solution
$$\cos 2x - \cos^2 x = 0$$
$$2\cos^2 x - 1 - \cos^2 x = 0$$
$$\cos^2 x - 1 = 0$$
$$\cos x = \pm 1$$
$$SS = \{0°,180°,360°\}$$

Check
$$\cos 2 \cdot 0° - \cos^2 0° = \cos 0° - \cos^2 0° = 1 - 1 = 0$$
$$\cos 2 \cdot 180° - \cos^2 180° = \cos 360° - 1^2 = 1 - 1 = 0$$
$$\cos 2 \cdot 360° - \cos^2 360° = \cos 720° - 1^2 = 1 - 1 = 0$$

Exercise 11-6.1 Solve the following equations. Use Table C-1 if approximations are necessary.

1 $\cos 2x + \cos x + 1 = 0$

2 $\sec 2x = 2 \sec x$

3 $\cos^2 2A = \cos^2 A$

4 $\cos 3B = \cos B$

5 $\sin 2A + 2 \sin^2 \frac{1}{2}A = 1$

6 $\tan 2A = -\cot A$

7 $\sin 2x + 2 = 4 \sin x + \cos x$

8 $\tan \frac{1}{2}A + 2 \sin 2A = \csc A$

9 $\cot \frac{1}{2}x - \cos x - 1 = 0$

10 $\dfrac{\sin x}{1 + \sin x} + \sin x = \dfrac{2}{\sin x}$

11 $\tan 2A = 2 - \cot 2A$

12 $\cos 2A + 2 \cos^2 \dfrac{A}{2} = 4$

Review Exercise Solve the following equations. Use Table C-1 to obtain approximations to the nearest degree.

1 (a) $\tan x = \sqrt{3}$

        (b) $\cot x = -\dfrac{1}{\sqrt{3}}$

    (c) $\cos x = -\frac{1}{2}$

        (d) $\tan x + 1 = 0$

    (e) $\sec x = -\sqrt{2}$

        (f) $\csc x - 2 = 0$

    (g) $\sin x = 5$

        (h) $\sin x = 0.2419$

2 (a) $2 \cos x - 1 = 0$

        (b) $\sqrt{3} \tan x + 1 = 0$

    (c) $2 \sec x - 1 = 0$

        (d) $2 \csc x - 4 = 0$

    (e) $5 - \sin x = 5.8660$

        (f) $\dfrac{4 \sec x - 3}{2 \sec x - 3} = \dfrac{8 \sec x + 5}{4 \sec x + 1}$

3 (a) $\sin^2 x = \sin x + 2$ (b) $2 \cos^2 x - 2 \sin^2 x = 1$

(c) $3 \cot^2 x = 1$ (d) $15 \cos^4 x - 22 \cos^2 x + 8 = 0$

(e) $\sin^2 x - 2 \sin x - 1 = 0$ (f) $\tan^2 x = 1 - 2 \tan x$

4 (a) $\sec^2 x + 4 \tan x = 0$ (b) $\sin x - \cos x = \sqrt{2}$

(c) $\sin^2 x - \cos^2 x = 0$ (d) $2 \sin^2 x + \sin x \cos x - \cos^2 x = 0$

(e) $\tan^2 x + \sec^2 x = 3$ (f) $\sec^2 x = 1 + \tan x$

5 (a) $\dfrac{\tan 2x + \tan x}{1 - \tan 2x \tan x} = \dfrac{1}{\sqrt{3}}$

(b) $\sin 2x + \sin x - 2 \cos x = 1$

(c) $\sin \tfrac{1}{2}x = \cos x$

(d) $2 \cos^2 2x - 3 \cos 2x + 1 = 0$

Answers

Exercise 11-2.1

1 SS $= \{60°, 300°\}$ **2** SS $= \{240°, 300°\}$

3 SS $= \{120°, 300°\}$ **4** SS $= \{60°, 240°\}$

5 SS $= \{63°, 243°\}$ **6** SS $= \{90°, 270°\}$

7 SS $= \{0°, 180°, 360°\}$ **8** SS $= \{135°, 315°\}$

9 SS $= \{120°, 240°\}$ **10** SS $= \{60°, 120°\}$

11 SS $= \{20°, 160°\}$ **12** SS $= \{110°, 290°\}$

Exercise 11-3.1	**1**	SS = {0°,360°}	**2** SS = {60°,300°}
	3	SS = {135°,315°}	**4** SS = {199°,341°}
	5	SS = {30°,150°}	**6** SS = {45°,225°}
	7	SS = {90°,270°}	**8** SS = {60°,300°}
	9	SS = {43°,137°}	**10** SS = {270°}

Exercise 11-4.1	**1**	SS = {55°,125°,235°,305°}	**2** SS = {60°,120°,240°,300°}
	3	SS = {30°,150°,210°,330°}	**4** SS = {45°,135°,225°,315°}
	5	SS = {0°,180°}	**6** SS = {0°,45°,180°,225°,360°}
	7	SS = {0°,71°,289°}	**8** SS = {23°,203°,142°,322°}
	9	SS = {90°}	**10** SS = ∅
	11	SS = ∅	**12** SS = {270°}
	13	SS = {90°}	**14** SS = {54°,234°,144°,324°}
	15	SS = {60°,120°,240°,300°}	**16** SS = {72°,252°,117°,297°}
	17	SS = {53°,127°,233°,307°}	**18** SS = {0°,60°,300°,360°}

Exercise 11-5.1	**1**	SS = {90°,210°,270°,330°}	**2** SS = {60°,120°,240°,300°,0°,360°,180°}
	3	SS = {0°,60°,180°,240°,360°}	**4** SS = {0°,180°,360°}
	5	SS = {135°,315°}	**6** SS = {53°,180°}
	7	SS = {71°,289°,180°}	**8** SS = {B|0° < B < 360°, B ≠ 180°}
	9	SS = {90°,270°,14°,194°}	**10** SS = {0°,60°,300°,360°}
	11	SS = ∅	**12** SS = {60°,300°}

Exercise 11-5.2	**1**	SS = ∅	**2** SS = {135°,315°}
	3	SS = {90°,180°}	**4** SS = {0°,360°,251°}
	5	SS = {270°,37°}	**6** SS = {15°,105°}

Exercise 11-6.1	**1**	SS = {90°,270°}	**2** SS = ∅
	3	SS = {0°,60°,120°,180°,240°,300°}	**4** SS = {0°,90°}
	5	SS = {90°,270°}	**6** SS = {90°,270°}
	7	SS = {30°,150°}	**8** SS = {90°,270°}
	9	SS = {90°,180°}	**10** SS = {224°,316°}
	11	SS = {23°,113°,203°,293°}	**12** SS = ∅

Review Exercise

1 (*a*) SS = {60°,240°} (*b*) SS = {120°,300°}
  (*c*) SS = {120°,240°} (*d*) SS = {135°,315°}
  (*e*) SS = {135°,225°} (*f*) SS = {30°,150°}
  (*g*) SS = ∅ (*h*) SS = {14°,166°}

2 (*a*) SS = {60°,300°} (*b*) SS = {150°,330°}
  (*c*) SS = ∅ (*d*) SS = {30°,150°}
  (*e*) SS = {240°,300°} (*f*) SS = {120°,240°}

3 (*a*) SS = {270°}
  (*b*) SS = {30°,150°,210°,330°}
  (*c*) SS = {60°,120°,240°,300°}
  (*d*) SS = {27°,153°,207°,333°,35°,145°,215°,325°}
  (*e*) SS = {204°,336°}
  (*f*) SS = {22°,202°,113°,293°}

4 (*a*) SS = {105°,165°,285°,345°} (*b*) SS = {135°}
  (*c*) SS = {45°,135°,225°,315°} (*d*) SS = {135°,315°}
  (*e*) SS = {45°,135°,225°,315} (*f*) SS = {0°,45°,180°,225°}

5 (*a*) SS = {10°,70°,190°,130°,250°,310°}
  (*b*) SS = {90°,120°,240°}
  (*c*) SS = {60°,300°}
  (*d*) SS = {0°,30°,150°,180°,210°,330°,360°}

Complex Numbers

12

Objectives

Upon completion of Chapter 12, you will be able to:

1 **Perform operations with complex numbers given in rectangular form. (12-2)**

2 **Graph complex numbers. (12-3)**

3 **Express in trigonometric form a complex number given in rectangular form. (12-4)**

4 **Express in rectangular form a complex number given in trigonometric form. (12-4)**

5 **Perform operations with complex numbers by using trigonometric form. (12-5)**

6 **Find powers of complex numbers. (12-6)**

7 **Find roots of complex numbers. (12-6)**

12/Complex Numbers

12-1 Introduction Perhaps you recall from your study of algebra that there are many equations which do not have solutions in the set of real numbers. For example,

$$x^2 + 1 = 0$$

and

$$2x^2 + x + 1 = 0$$

do not have real number solutions. Since the square of each real number is never negative, the need for a set of numbers which provides us with the number x such that $x^2 = -1$ is apparent.

In this chapter we will define this set of numbers and describe a few of the relations, operations, and properties of the set. The development of this set of numbers has initiated the study of a large part of mathematics, and it has enhanced the application of mathematics in engineering, the physical sciences, and other areas.

12-2 Complex Numbers A number of the form $a + bi$, where a and b are real numbers and $i^2 = -1$, is called a *complex number*. Hence the set of complex numbers, denoted by C, is given by

$$C = \{x \mid x = a + bi, a, b \in R, i^2 = -1\}$$

where a is called the *real component* and b is called the *imaginary component*, of the number $a + bi$.

Two complex numbers $a + bi$ and $c + di$ are equal if and only if $a = c$ and $b = d$. The operations of addition, multiplication, subtraction, and division on C are given by the following equations:

$$(a + bi) + (c + di) = (a + c) + (b + d)i$$

$$(a + bi)(c + di) = (ac - bd) + (bc + ad)i$$

$$(a + bi) - (c + di) = (a - c) + (b - d)i$$

$$\frac{a + bi}{c + di} = \frac{(ac + bd) + (bc - ad)i}{c^2 + d^2}$$

The closure, commutative, associative, and distributive laws also hold true for the set of complex numbers.

Two complex numbers $a + bi$ and $a - bi$ are said to be *conjugate* complex numbers. Each number is the conjugate of the other. Note that when two conjugate complex numbers are multiplied, the product is a real number. It is interesting to note that i raised to any integral power greater than 1 will always be i, -1, $-i$, or 1. That is,

$$i^2 = -1$$

$$i^3 = i^2 i = -1 \cdot i = -i$$

$$i^4 = i^2 i^2 = (-1)(-1) = 1$$

The value of i to any integral power m greater than 4 can be found by dividing m by 4 and using the remainder r as a power of i in the product $i^{m/4} \cdot i^r$. For example, the value of i^{219} can be found by

$$i^{219} = i^{216} \cdot i^3$$

$$= (i^4)^{54} \cdot i^3$$

$$= 1^{54} \cdot i^3$$

$$= 1 \cdot i^3$$

$$= -i$$

The following examples illustrate some operations and properties of the complex number system.

Example 12-2.1 Find the sum and product of $(2 - 5i)$ and $(-3 - 2i)$.

Solution
$$(2 - 5i) + (-3 - 2i) = [2 + (-3)] + [-5 + (-2)]i$$
$$= -1 - 7i$$

$$(2 - 5i)(-3 - 2i) = [2(-3) - (-5)(-2)] + [(-5)(-3) + 2(-2)]i$$
$$= (-6 - 10) + (15 - 4)i$$
$$= -16 + 11i$$

Example 12-2.2 Simplify $i^{10} - 3i^8 + 2i^{15}$.

Solution Since $i^4 = i^2 \cdot i^2 = (-1)(-1) = 1$, i with an exponent which is an integral multiple of 4 is always 1. Also, integral powers of i will always yield i, -1, $-i$, or 1.

$$i^{10} - 3i^8 + 2i^{15} = i^8(i^2 - 3 + 2i^7) = 1(-1 - 3 + 2i^4 \cdot i^3)$$
$$= -4 + 2i^3 = -4 - 2i$$

Example 12-2.3 Solve the equation $8a + (a - 3b)i = 2b + 4i$ for a and b.

Solution If $8a + (a - 3b)i = 2b + 4i$, then by the definition of equality of complex numbers

$$8a = 2b \quad \text{and} \quad a - 3b = 4$$

$$4a = b \quad \text{and} \quad a - 3(4a) = 4$$

Hence,

$$-11a = 4 \quad a = -\tfrac{4}{11} \quad \text{and} \quad b = -\tfrac{16}{11}$$

Example 12-2.4 Perform the indicated division and leave the result in the form $a + bi$:

$$\frac{3 - 2i}{5 + i}$$

Solution From the definition of division of complex numbers it follows that

$$\frac{3 - 2i}{5 + i} = \frac{[3 \cdot 5 + (-2)1] + [(-2) \cdot 5 - 3 \cdot 1]i}{5^2 + 1^2}$$

$$= \frac{13 - 13i}{26} = \tfrac{1}{2} - \tfrac{1}{2}i$$

An alternative method of solution employs the conjugate of the denominator.

$$\frac{3 - 2i}{5 + i} = \frac{3 - 2i}{5 + i} \cdot \frac{5 - i}{5 - i} = \frac{3 \cdot 5 - 3i - 10i + 2i^2}{25 - i^2} = \frac{13 - 13i}{26} = \tfrac{1}{2} - \tfrac{1}{2}i$$

Exercise 12-2.1 **1** Simplify the following expressions:

(a) $3i^3 - 2i^2 + i$ (b) $6i^4 + 2i^8 - i^{10}$

(c) i^{50} (d) $i^6 - i^5 - i^4 - i^3 - i^2 - i$

2 Solve each equation for a and b.

(a) $3 + 2i = a + bi$ (b) $a - 2bi = 7 - 3i$

(c) $a + 3b - 2i = 2 + (4b - a)i$ (d) $5a + (a + b)i = 3b + 6i$

3 Find the conjugate of each of the following:

(a) $\sqrt{5} - \sqrt{5}i$ (b) $(2 + i) + (3 - i)$

(c) $2 - \sqrt{7}i$ (d) $(4 + \sqrt{3}i) + (2\sqrt{2} - i)$

4 Perform the indicated operations and leave the answer in the form $a + bi$.

(a) $(2 + 3i)^2$ (b) $(3 - i) + (7 - 2i)$

(c) $(2 + i) + (5 - 3i) - (3 - 2i)$ (d) $\dfrac{1 + i}{2 + i} - \dfrac{3i}{4 + i}$

(e) $(5 + 3i) \div (7 - i)$ (f) $\left(\dfrac{1 + 2i}{5 - i}\right)\left(\dfrac{3i}{2 + i}\right)$

(g) $(2 + i) + \dfrac{5 - 11i}{2 + 3i}$ (h) $\dfrac{5}{4 + 3i}$

(i) $(1 + i)(2 + i)(1 - i)$

12-3 Graphing Complex Numbers   Since the complex number $a + bi$ has the real numbers a and b as real and imaginary components, respectively, we can associate each complex number $a + bi$ with the unique ordered pair of real numbers (a,b). Hence, to graph the complex number $a + bi$, we merely select two perpendicular axes, just as we did with the rectangular coordinate system, and represent the elements of C as points of the complex plane. Note in Figure 12-3.1 that the horizontal axis is called the real axis and the vertical axis is called the imaginary axis. Note further that the real component of $a + bi$ is plotted on the horizontal axis and the imaginary component of $a + bi$ is plotted on the vertical axis.

Figure 12-3.1

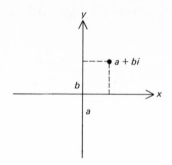

Example 12-3.1 Plot the following complex numbers: $3i$, 5, $2 + 2i$, $4 - i$, and $-2 - 3i$.

Solution

Figure 12-3.2

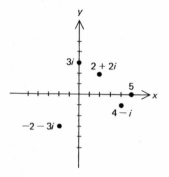

Figure 12-3.3 indicates how addition of complex numbers is accomplished graphically. This idea will have additional significance when you study vectors.

Figure 12-3.3

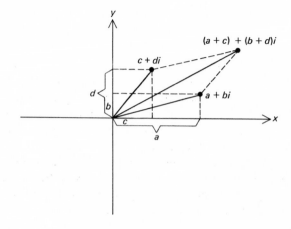

Exercise 12-3.1   **1**   Plot the following numbers in the complex plane.

(a) $\sqrt{3} + i$ (b) $\sqrt{17}$

(c) $-\sqrt{5} - 2i$ (d) $-\sqrt{7} + \sqrt{7}i$

(e) $3\sqrt{2} + i$ (f) $(2 + i) + (-3 - 2i)$

(g) 0 (h) $-2i$

(i) $4 + 2i$ (j) $5 - 7i$

(k) $-3 - \frac{1}{2}i$ (l) $\frac{5}{7} + \frac{4}{5}i$

12-4 Trigonometric Form of Complex Numbers
In the last section we observed that the complex number $a + bi$ is plotted in the complex plane as the point (a,b).   The distance from the origin to the point (a,b) is called the *modulus* of the complex number $a + bi$, and it is denoted by $|a + bi|$.   In Figure 12-4.1 we see from the pythagorean

Figure 12-4.1

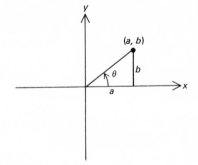

theorem that $|a + bi| = \sqrt{a^2 + b^2}$.   We also observe that if θ is an angle in standard position with the line from $(0,0)$ to (a,b) as its terminal side and $r = |a + bi|$, then from the definitions of the sine and cosine functions,

$$\cos \theta = \frac{a}{r} \qquad \sin \theta = \frac{b}{r}$$

$$a = r \cos \theta \qquad b = r \sin \theta$$

$$\tan \theta = \frac{b}{a}$$

Hence, $a + bi = r \cos \theta + (r \sin \theta)i = r(\cos \theta + i \sin \theta)$. The expression $r(\cos \theta + i \sin \theta)$ is called the *trigonometric form*, or *polar form*, for the complex number $a + bi$. The form $a + bi$ is called the *rectangular form*, or *algebraic form*, of the number.

Since the sine and cosine functions both have a cycle of 360°, it is readily seen that for any integer n

$$a + bi = r[\cos (\theta + n \cdot 360°) + i \sin (\theta + n \cdot 360°)]$$

Thus there are infinitely many expressions for $a + bi$ in trigonometric form. The least positive value for θ is most frequently used.

The following examples indicate how we change a complex number from rectangular form to trigonometric form and vice versa.

Example 12-4.1   Express $2 + i$ in trigonometric form.

Solution   Since $a = 2$ and $b = 1$, $r = \sqrt{a^2 + b^2} = \sqrt{2^2 + 1^2} = \sqrt{5}$. $\tan \theta = \frac{b}{a} = \frac{1}{2}$, so $\theta = 27°$ from Table C-1. Hence, $2 + i = \sqrt{5}(\cos 27° + i \sin 27°)$.

Example 12-4.2   Express $\sqrt{2}(\cos 315° + i \sin 315°)$ in rectangular form, that is, $a + bi$.

Solution

$$r = \sqrt{2} \qquad\qquad \theta = 315°$$

$$a = r \cos \theta \qquad\qquad b = r \sin \theta$$

$$a = \sqrt{2} \cos 315° \qquad\quad b = \sqrt{2} \sin 315°$$

$$a = \sqrt{2} \cdot \frac{1}{\sqrt{2}} = 1 \qquad b = \sqrt{2} \cdot \left(-\frac{1}{\sqrt{2}}\right) = -1$$

Hence, the rectangular form is $1 - i$.

Exercise 12-4.1   **1**   Express each of the following complex numbers in trigonometric form, using the angle with least positive measure. Use Table C-1 if approximations are necessary.

(a) $2 - 2i$ (b) $-i$

(c) $\sqrt{2} - i\sqrt{2}$ (d) $-1 - i\sqrt{3}$

(e) $4 - 3i$ (f) $\sqrt{7} + i\sqrt{21}$

(g) $-6 + 6i\sqrt{3}$ (h) $-3i$

(i) 5   (j) $(1 + i)^2$

2 Express each of the following numbers in rectangular form:

(a) $\cos 45° + i \sin 45°$ (b) $\cos 12° + i \sin 12°$

(c) $2(\cos 180° + i \sin 180°)$ (d) $3[\cos (-45°) + i \sin (-45°)]$

(e) $\sqrt{3}(\cos 300° + i \sin 300°)$ (f) $5(\cos 38°20' + i \sin 38°20')$

(g) $4(\cos 315° + i \sin 315°)$ (h) $\cos 135° + i \sin 135°$

(i) $6(\cos 270° + i \sin 270°)$ (j) $2(\cos 120° + i \sin 120°)$

12-5 Multiplication and Division of Complex Numbers in Trigonometric Form

One great advantage of having complex numbers in trigonometric form is the simplicity of multiplication and division in this form. The properties for multiplication and division are given below.

$$r_1(\cos \theta_1 + i \sin \theta_1) \cdot r_2(\cos \theta_2 + i \sin \theta_2) = r_1 r_2[\cos (\theta_1 + \theta_2) + i \sin (\theta_1 + \theta_2)]$$

$$\frac{r_1(\cos \theta_1 + i \sin \theta_1)}{r_2(\cos \theta_2 + i \sin \theta_2)} = \frac{r_1}{r_2}[\cos (\theta_1 - \theta_2) + i \sin (\theta_1 - \theta_2)]$$

The proofs of these properties are left as an exercise.

Example 12-5.1 Express $1 + i$ and $1 + i\sqrt{3}$ in trigonometric form and find their product.

Solution
$$1 + i = \sqrt{2}(\cos 45° + i \sin 45°)$$
$$1 + i\sqrt{3} = 2(\cos 60° + i \sin 60°)$$
$$(1 + i)(1 + i\sqrt{3}) = [\sqrt{2}(\cos 45° + i \sin 45°)][2(\cos 60° + i \sin 60°)]$$
$$= 2\sqrt{2}(\cos 105° + i \sin 105°)$$

Example 12-5.2 Use the trigonometric form to find $\dfrac{3 - i}{1 + 2i}$.

Solution
$$3 - i = \sqrt{10}(\cos 343° + i \sin 343°)$$
since $a = 3$, $b = -1$, $\tan \theta = -\tfrac{1}{3}$, and $\theta = 343°$.

$$1 + 2i = \sqrt{5}(\cos 63° + i \sin 63°)$$

$$\frac{3 - i}{1 + 2i} = \frac{\sqrt{10}(\cos 343° + i \sin 343°)}{\sqrt{5}(\cos 63° + i \sin 63°)}$$

$$= \frac{\sqrt{10}}{\sqrt{5}}[\cos (343° - 63°) + i \sin (343° - 63°)]$$

$$= \sqrt{2}(\cos 280° + i \sin 280°)$$

Exercise 12-5.1 **1** Perform the indicated operations by using the trigonometric form.

(a) $(1 + i)(2 - 3i)$

(b) $(7 - \sqrt{2}i)(1 - \sqrt{3}i)$

(c) $(2 + 2i)(2 - 2\sqrt{3}i)$

(d) $(5 - \sqrt{5}i)(5 + \sqrt{5}i)$

(e) $(\sqrt{3} - i)(1 + i)$

(f) $\dfrac{2\sqrt{3} + 2i}{1 + i\sqrt{3}}$

(g) $\dfrac{3 + i}{4 + i}$

(h) $\dfrac{1 + i}{1 - i\sqrt{3}}$

(i) $\dfrac{2}{3 - i}$

(j) $\dfrac{-1 - 2i}{2 + 4i}$

2   Prove the multiplication property of complex numbers in trigonometric form.

3   Prove the division property of complex numbers in trigonometric form.

12-6   Powers and Roots of Complex Numbers

It is often necessary to find positive integral powers of complex numbers and roots of complex numbers. The following property (known as De Moivre's theorem) helps us to perform these tasks.

$$[r(\cos \theta + i \sin \theta)]^n = r^n(\cos n\theta + i \sin n\theta) \qquad n \text{ a real number}$$

The proof of this theorem is beyond the scope of this book; however, the following examples illustrate its use.

Example 12-6.1   Find the value of $[2(\cos 30° + i \sin 30°)]^4$.

Solution
$$[2(\cos 30° + i \sin 30°)]^4 = 2^4(\cos 120° + i \sin 120°)$$
$$= 16\left(-\frac{1}{2} + i \cdot \frac{\sqrt{3}}{2}\right)$$
$$= -8 + 8i\sqrt{3}$$

Example 12-6.2   Find the value of $(1 + i)^5$.

Solution
$$(1 + i)^5 = [\sqrt{2}(\cos 45° + i \sin 45°)]^5$$
$$= (\sqrt{2})^5(\cos 225° + i \sin 225°)$$
$$= 4\sqrt{2}\left[-\frac{1}{\sqrt{2}} + i\left(-\frac{1}{\sqrt{2}}\right)\right]$$
$$= -4 - 4i$$

You learned in algebra that if m is a positive real number and n is a positive integer greater than 1, then there is exactly one positive real number a such that $a^n = m$. This number a is called the nth root of m and is denoted by $\sqrt[n]{m}$. The complex number system is different in that any nonzero complex number has n nth roots. The following property is used to find roots of complex numbers.

A nonzero number $r(\cos \theta + i \sin \theta)$ has n nth roots which are given by

$$\sqrt[n]{r}\left(\cos \frac{\theta + 2\pi k}{n} + i \sin \frac{\theta + 2\pi k}{n}\right) \qquad k = 0, 1, \ldots, n - 1$$

Example 12-6.3 Find the cube roots of $27(\cos 60° + i \sin 60°)$.

Solution The three cube roots are

$$\sqrt[3]{27}\left(\cos \frac{60° + k \cdot 360°}{3} + i \sin \frac{60° + k \cdot 360°}{3}\right) \qquad k = 0, 1, 2$$

For $k = 0$, we get $3(\cos 20° + i \sin 20°)$. For $k = 1$, we get $3(\cos 140° + i \sin 140°)$. For $k = 2$, we get $3(\cos 260° + i \sin 260°)$.

Example 12-6.4 Find the fourth roots of $(-8i)$.

Solution $\qquad -8i = 0 - 8i \qquad a = 0 \qquad b = -8$

Hence, $\theta = 270°$ and $r = 8$.

$$-8i = 8(\cos 270° + i \sin 270°)$$

The four fourth roots are given by

$$\sqrt[4]{8}\left(\cos \frac{270° + k(360°)}{4} + i \sin \frac{270° + k(360°)}{4}\right)$$

Substitution of the values for k yields the following values:

For $k = 0$, $\qquad \sqrt[4]{8}(\cos 67.5° + i \sin 67.5°)$

For $k = 1$, $\qquad \sqrt[4]{8}(\cos 157.5° + i \sin 157.5°)$

For $k = 2$, $\qquad \sqrt[4]{8}(\cos 247.5° + i \sin 247.5°)$

For $k = 3$, $\qquad \sqrt[4]{8}(\cos 337.5° + i \sin 337.5°)$

Exercise 12-6.1 **1** Find the indicated powers and give the answers in rectangular form.

(a) $[2(\cos 20° + i \sin 20°)]^3$ $\qquad$ (b) $[4(\cos 15° + i \sin 15°)]^3$

(c) $(\cos 150° + i \sin 150°)^4$ $\qquad$ (d) $(1 - i\sqrt{3})^5$

(e) $(\cos 210° + i \sin 210°)^6$ (f) $(1 + i)^4$

(g) $(-2 - 2i)^5$ (h) $[2(\cos 45° + i \sin 45°)]^6$

2   Find the indicated roots and leave the answers in trigonometric form.

(a)   The fourth roots of $16(\cos 60° + i \sin 60°)$

(b)   The cube roots of $8(\cos 240° + i \sin 240°)$

(c)   The square roots of $3 - 4i$

(d)   The fifth roots of $16 - 16i\sqrt{3}$

(e)   The cube roots of $1 - i$

Review Exercise   **1**   Perform the indicated operations and leave the answer in the form $a + bi$.

(a) $(-\sqrt{2} + i\sqrt{2})(3 - i)$ (b) $(2 - i)^3$

(c) $(-5 + 2i) + (-3 - 2i)$ (d) $(2 + 4i)(1 + i)(3 - i)$

(e) $(2 - i) - (\sqrt{5} + 3i)$ (f) $\dfrac{1 - i}{1 + i}$

(g) i^{50} (h) $\dfrac{-6 - 2i}{-3 + i}$

2   Plot the following numbers in the complex plane:

(a)   4   (b) $4 - i$

(c) $2 + 7i$

(d) $\sqrt{7} - \sqrt{3}i$

(e) $-\sqrt{5} + \frac{1}{2}i$

(f) $\frac{2}{3} - \sqrt{13}i$

3 Express each of the following complex numbers in trigonometric form, using the angle with least positive measure. Use Table C-1 for approximations.

(a) $\sqrt{17} + i\sqrt{2}$

(b) $3 - 3i$

(c) $1 - i\sqrt{3}$

(d) $\frac{5}{2} + \frac{5}{2}i$

(e) $2 - 9i$

(f) $-8 - 2i$

(g) $3\sqrt{3} - 3i$

(h) -5

4 Express each of the following complex numbers in rectangular form:

(a) $2(\cos 60° + i \sin 60°)$

(b) $4(\cos 225° + i \sin 225°)$

(c) $\frac{1}{4}(\cos 210° + i \sin 210°)$

(d) $2(\cos 300° + i \sin 300°)$

(e) $3(\cos 60° + i \sin 60°)$

5 Perform the indicated operations, using trigonometric form, and leave the answer in trigonometric form.

(a) $(-2 + 2i\sqrt{3})(5 + i)$ (b) $(\sqrt{3} - i)(7 + i)$

(c) $\dfrac{1 - i\sqrt{3}}{1 - i}$

(d) $(\cos 15° + i \sin 15°)(\cos 10° + i \sin 10°)$

(e) $5(\cos 20° + i \sin 20°) \cdot 3(\cos 25° + i \sin 25°)$

6 Find the indicated powers and give the answers in trigonometric form.

(a) $[4(\cos 20° + i \sin 20°)]^3$ (b) $\left(\dfrac{1}{3} + i\dfrac{\sqrt{3}}{3}\right)^3$

(c) $(\cos 180° + i \sin 180°)^5$ (d) $(2i)^6$

(e) $(-3 - i)^5$ (f) $(\sin 60° + i \cos 60°)^2$

7 Find the indicated roots and leave the answers in trigonometric form.

(a) The fourth roots of $81(\cos 300° + i \sin 300°)$

(b) The fifth roots of -32

(c) The cube roots of $-3 - 4i$

(d) The cube roots of $-8i$

(e) The square roots of $4(\cos \frac{2}{3}\pi + i \sin \frac{2}{3}\pi)$

(f) The fourth roots of $16\left(\cos \dfrac{7\pi}{4} + i \sin \dfrac{7\pi}{4}\right)$

Answers

Exercise 12-2.1   **1** (a) $2 - 2i$ (b) 9 (c) -1 (d) $-1 - i$
   2 (a) $3 = a, 2 = b$ (b) $a = 7, b = \frac{3}{2}$
       (c) $b = 0, a = 2$ (d) $a = \frac{9}{4}, b = \frac{15}{4}$
   3 (a) $\sqrt{5} + \sqrt{5}i$ (b) 5
       (c) $2 + \sqrt{7}i$ (d) $(4 + 2\sqrt{2}) - (\sqrt{3} - 1)i$
   4 (a) $-5 + 12i$ (b) $10 - 3i$ (c) 4 (d) $\frac{36}{85} - \frac{43}{85}i$
       (e) $\frac{16}{25} + \frac{13}{25}i$ (f) $-\frac{57}{130} + \frac{51}{130}i$ (g) $\frac{3}{13} - \frac{24}{13}i$
       (h) $\frac{4}{5} - \frac{3}{5}i$ (i) $4 + 2i$

(e) $\frac{16}{25} + \frac{13}{25}i$ (f) $\frac{57}{130} + \frac{51}{130}i$ (g) $\frac{3}{13} - \frac{24}{13}i$
(h) $\frac{4}{5} - \frac{3}{5}i$ (i) $4 + 2i$

Exercise 12-3.1   **1**

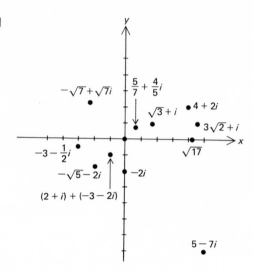

Exercise 12-4.1   **1**   (a) $2\sqrt{2}(\cos 315° + i \sin 315°)$ (b) $(\cos 270° + i \sin 270°)$
(c) $2(\cos 315° + i \sin 315°)$ (d) $2(\cos 240° + i \sin 240°)$
(e) $5(\cos 323° + i \sin 323°)$ (f) $r = 2\sqrt{7}(\cos 60° + i \sin 60°)$
(g) $12(\cos 120° + i \sin 120°)$ (h) $3(\cos 270° + i \sin 270°)$
(i) $5(\cos 0° + i \sin 0°)$ (j) $2(\cos 90° + i \sin 90°)$

2   (a) $\sqrt{2} + i\sqrt{2}$ (b) $0.9781 + 0.2079i$
(c) -2 (d) $\frac{3}{2}\sqrt{2} - \frac{3}{2}\sqrt{2}i$
(e) $\dfrac{\sqrt{3}}{2} - \dfrac{3}{2}i$ (f) $3.922 + 3.101i$
(g) $2\sqrt{2} - 2i\sqrt{2}$ (h) $-\dfrac{\sqrt{2}}{2} + \dfrac{i\sqrt{2}}{2}$
(i) $-6i$ (j) $-1 + i\sqrt{3}$

Exercise 12-5.1   **1**   (a) $5 - i$ (b) $(7 - \sqrt{6}) - (\sqrt{2} + 7\sqrt{3})i$
(c) $(4 + 4\sqrt{3}) + (4 - 4\sqrt{3})i$ (d) 30
(e) $(\sqrt{3} + 1) + (\sqrt{3} - 1)i$ (f) $\sqrt{3} - i$
(g) $\frac{13}{17} + \frac{1}{17}i$ (h) $\dfrac{1 - \sqrt{3}}{4} + \dfrac{1 + \sqrt{3}}{4}i$
(i) $\frac{3}{5} + \frac{1}{5}i$ (j) $-\frac{1}{2}$

2   *Proof:* $r_1(\cos \theta_1 + i \sin \theta_1) \cdot r_2(\cos \theta_2 + i \sin \theta_2)$
$= r_1 r_2 (\cos \theta_1 + i \sin \theta_1)(\cos \theta_2 + i \sin \theta_2)$
$= r_1 r_2 [(\cos \theta_1 \cos \theta_2 - \sin \theta_1 \sin \theta_2)$
$\qquad\qquad + i(\sin \theta_1 \cos \theta_2 + \cos \theta_1 \sin \theta_2)]$
$= r_1 r_2 [\cos (\theta_1 + \theta_2) + i \sin (\theta_1 + \theta_2)]$

3 *Proof:* $= \dfrac{r_1(\cos \theta_1 + i \sin \theta_1)}{r_2(\cos \theta_2 + i \sin \theta_2)} \cdot \dfrac{r_2[\cos (-\theta_2) + i \sin (-\theta_2)]}{r_2[\cos (-\theta_2) + i \sin (-\theta_2)]}$

$= \dfrac{r_1 r_2[\cos (\theta_1 - \theta_2) + i \sin (\theta_1 - \theta_2)]}{r_2{}^2(\cos \theta + i \sin \theta)}$

$= \dfrac{r_1}{r_2}[\cos (\theta_1 - \theta_2) + i \sin (\theta_1 - \theta_2)]$

Exercise 12-6.1

1   (*a*) $4 + 4i\sqrt{3}$ (*b*) $32\sqrt{2} + 32i\sqrt{2}$

     (*c*) $-\dfrac{1}{2} - \dfrac{i\sqrt{3}}{2}$ (*d*) $16 + 16i\sqrt{3}$

     (*e*) -1 (*f*) -4

     (*g*) $128 + 128i$ (*h*) $-64i$

2   (*a*) $2(\cos 15° + i \sin 15°)$ $2(\cos 195° + i \sin 195°)$

           $2(\cos 105° + i \sin 105°)$ $2(\cos 285° + i \sin 285°)$

     (*b*) $2(\cos 80° + i \sin 80°)$ $2(\cos 320° + i \sin 320°)$

           $2(\cos 200° + i \sin 200°)$

     (*c*) $\sqrt{5}(\cos 154° + i \sin 154°)$ $\sqrt{5}(\cos 334° + i \sin 334°)$

     (*d*) $\sqrt[5]{26}\sqrt[10]{2}\left(\dfrac{\cos 300° + k \cdot 360°}{5} + i \sin \dfrac{300° + k \cdot 360°}{5}\right)$

                                     for $k = 0, 1, 2, 3, 4$

     (*e*) $\sqrt[6]{2}(\cos 108° + i \sin 108°)$

           $\sqrt[6]{2}(\cos 228° + i \sin 228°)$

           $\sqrt[6]{2}(\cos 348° + i \sin 348°)$

Review Exercise

1   (*a*) $-2\sqrt{2} + 4\sqrt{2}i$ (*b*) $2 - 11i$

     (*c*) -8 (*d*) $20i$

     (*e*) $(2 - \sqrt{5}) + (-4i)$ (*f*) $-i$

     (*g*) -1 (*h*) $\dfrac{8}{5} + \dfrac{6i}{5}$

2

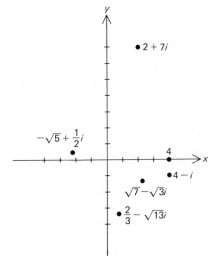

3 (a) $\sqrt{19}(\cos 18°55' + i \sin 18°55')$ (b) $3\sqrt{2}(\cos 315° + i \sin 315°)$
(c) $2(\cos 300° + i \sin 300°)$ (d) $\frac{5}{2}\sqrt{2}(\cos 45° + i \sin 45°)$
(e) $\sqrt{85}(\cos 282°32' + i \sin 282°32')$ (f) $2\sqrt{17}(\cos 194° + i \sin 194°)$
(g) $6(\cos 330° + i \sin 330°)$ (h) $5(\cos 180° + i \sin 180°)$

4 (a) $1 + i\sqrt{3}$ (b) $-2\sqrt{2} - 2\sqrt{2}i\dfrac{i\sqrt{2}}{2}$

(c) $-\dfrac{\sqrt{3}}{8} + \dfrac{i}{8}$ (d) $1 - i\sqrt{3}$

(e) $\dfrac{3}{2} + \dfrac{3i\sqrt{3}}{2}$

5 (a) $4\sqrt{26}(\cos 131°21' + i \sin 131°21')$
(b) $10\sqrt{2}(\cos 338°8' + i \sin 338°8')$
(c) $\sqrt{2}(\cos 165° + i \sin 165°)$
(d) $(\cos 25° + i \sin 25°)$
(e) $15(\cos 45° + i \sin 45°)$

6 (a) $64(\cos 45° + i \sin 45°)$ (b) $\frac{8}{27}(\cos 180° + i \sin 180°)$
(c) $\cos 900° + i \sin 900°)$ (d) $64(\cos 540° + i \sin 540°)$
(e) $10^{5/2}(\cos 272°10' + i \sin 272°10')$ (f) $\cos 120° + i \sin 120°$

7 (a) $3(\cos 75° + i \sin 75°)$ $3(\cos 165° + i \sin 165°)$
     $3(\cos 255° + i \sin 255°)$ $3(\cos 345° + i \sin 345°)$
(b) $2(\cos 36° + i \sin 36°)$ $2(\cos 108° + i \sin 108°)$
     $2(\cos 180° + i \sin 180°)$ $2(\cos 252° + i \sin 252°)$
     $2(\cos 324° + i \sin 324°)$
(c) $\sqrt[3]{5}(\cos 77.72° + i \sin 77.72°)$ $\sqrt[3]{5}(\cos 197.06° + i \sin 197.06°)$
     $\sqrt[3]{5}(\cos 317.72° + i \sin 317.72°)$
(d) $2(\cos 90° + i \sin 90°)$ $2(\cos 210° + i \sin 210°)$
     $2(\cos 330° + i \sin 330°)$
(e) $2\left(\cos \dfrac{\pi}{3} + i \sin \dfrac{\pi}{3}\right)$ $2\left(\cos \dfrac{4\pi}{3} + i \sin \dfrac{4\pi}{3}\right)$
(f) $2\left(\cos \dfrac{7\pi}{16} + i \sin \dfrac{7\pi}{16}\right)$ $2\left(\cos \dfrac{15\pi}{16} + i \sin \dfrac{15\pi}{16}\right)$
     $2\left(\cos \dfrac{23}{16}\pi + i \sin \dfrac{23}{16}\pi\right)$ $2\left(\cos \dfrac{31}{16}\pi + i \sin \dfrac{31}{16}\pi\right)$

Appendix Basic Algebra Review

Objectives

Upon completion of Appendix A, you will be able to perform the following:

1 **Describe a set by the roster method. (A-2)**

2 **Describe a set by the set builder notation method. (A-2)**

3 **Write the cartesian product of two sets. (A-2)**

4 **Determine whether a given equation defines a relation. (A-3)**

5 **Determine whether a given equation defines a function. (A-3)**

6 **Write the domain of a given relation. (A-3)**

7 **Write the range of a given relation. (A-3)**

8 **Sketch the graphs of selected algebraic functions. (A-4)**

A/Basic Algebra Review

A-1 Introduction These topics facilitate understanding of the relationship between algebra and trigonometry and help explain the structure of trigonometry.

A-2 Sets We will assume that a set is any collection of elements or objects. An example of a set is the set of counting numbers less than twenty-five. Capital letters and braces are commonly used to represent sets. The above example could be represented as $C = \{1, 2, 3, \ldots, 24\}$. This method of describing a set by listing the elements inside braces is called the *roster method*, or *listing method*. Below are three additional examples which illustrate the roster method.

Example A-2.1 $A = \{2,4,6\}$

Example A-2.2 $B = \{(x,y,z), (1,2,3), (\alpha,\theta,\gamma)\}$

Example A-2.3 $C = \{\ldots,-4,-2,-,2,4,\ldots\}$

Note that this method may be used to represent either an infinite set or a finite set.[1]

In this book the following capital letters are used to denote some special set of numbers:

$N = \{\text{natural numbers}\} = \{1,2,3,\ldots\}$

$Z = \{\text{integers}\} = \{\ldots,-3,-2,-1,0,1,2,3,\ldots\}$

$Q = \{\text{rational numbers}\} = \{\text{infinite repeating or terminating decimals}\}$

$H = \{\text{irrational numbers}\} = \{\text{infinite nonrepeating decimals}\}$

$R = \{\text{real numbers}\} = Q \cup H$

Exercise A-2.1 Use the roster method to describe the following:

1 The set of even natural numbers less than or equal to 12

---

[1]In this text a finite set will denote a set which can be placed in one-to-one correspondence with the set $\{1,2,3,\ldots,n\}$ for some fixed natural number n. A set which is not finite is called *infinite*.

2 The set of positive integral multiples of 3

3 The set of odd integers between 0 and 16

4 The set of positive integers less than 7 and greater than or equal to 15

5 The set of values for y where $y = (\frac{1}{2})^x$ and $x = 0, -1, -2$

6 The set of values for k where $k + r = 5$ and $r = 0, -1, 2$

Another method commonly used to describe a set is *set builder notation*. This method specifies a variable and then describes a condition on the variable. Set builder notation is illustrated by the following examples.

Example A-2.4 $D = \{x|x$ is an even integer$\}$. Read "the set of all elements x such that x is an even integer."

Example A-2.5 $M = \{y|y = 2x^2, x$ is any real number$\}$. Read "the set of all elements y such that $y = 2x^2$ for any real number x."

Example A-2.6 $K = \{(x,y,z)|x, y, z$ are any real numbers$\}$. Read "the set of all ordered triples (x,y,z) such that x, y, z are any real numbers."

Example A-2.7   The set of rational numbers may be written $Q = \left\{\dfrac{a}{b}\middle| a \text{ and } b \text{ are elements}\right.$ of Z and $b \neq 0\}$.

Exercise A-2.2   Describe the following sets by using set builder notation:

1 The set of real numbers between $\sqrt{3}$ and $\frac{3}{11}$

2 The set of letters in the alphabet between f and m

3 The set of all ordered pairs (x,y) such that $y = \sqrt{2x^2 + 3}$

4 The set of positive odd integral multiples of 3

5 The set of positive integral multiples of 4

6 The set of all angles between $0°$ and $90°$

The symbol $\in$ indicates "belongs to" or "is an element of." In reference to Example A-2.1, we write $2 \in A$ to denote that 2 is an element of set A.

The *cartesian product* between sets A and B, denoted $A \times B$, is defined as follows:

$A \times B = \{(a,b)|a \in A, b \in B\}$ is the set of ordered pairs (a,b) such that a belongs to set A and b belongs to set B. Note the importance of the word "ordered." The ordered pair (a,b) is not the same as (b,a). Hence, the ordered pair $(a,b) \in A \times B$ while $(b,a) \in B \times A$.

If A contains n elements and B contains m elements, then $A \times B$ will contain $n \times m$ elements, since each element in A must be used with each element in B to form the entire set $A \times B$. Thus, in example A-2.8, $A \times B$ contains 2×2, or 4, elements; in example A-2.9, $A \times B$ contains 3×2, or 6, elements; in example A-2.10, $A \times A$ contains 2×2, or 4, elements; and in example A-2.11, $N \times N$ contains infinitely many elements.

Example A-2.8

$A = \{1,5\}, B = \{2,3\}$

$A \times B = \{(1,2), (1,3), (5,2), (5,3)\}$

Example A-2.9

$A = \{x,y,z\}, B = \{a,x\}$

$A \times B = \{(x,a), (x,x), (y, a), (y,x), (z,a), (z,x)\}$

Example A-2.10

$A = \{k,r\}$

$A \times A = \{(k,k), (k,r), (r,k), (r,r)\}$

Example A-2.11 $N = \{x|x \text{ is a natural number}\}$

$N \times N = \{(x,y)|x, y \in N\}$

Exercise A-2.3   **1** If $A = \{p,q,r\}$ and $B = \{s,t\}$, write $A \times B$. How many elements does $A \times B$ contain?

2 If $C = \{30°,60°\}$ and $M = \{\frac{1}{2},\sqrt{3}\}$, write $C \times M$ and $M \times C$. Is $M \times C = C \times M$? How many elements are in each?

3 If $R = \{x|x \text{ is a real number}\}$, describe $R \times R$.

4 Show that if A is a finite set, the number of elements in $A \times A$ is a perfect square.

If A is a set, then B is a *subset* of A if each element of B is an element of A. The symbol $\subset$ denotes "a subset of," and we write $B \subset A$ to mean B is a subset of A.

Example A-2.12   If $P = \{(x,y)|y = 3x, x \text{ is a real number}\}$ and $A = \{(1,3),(-1,-3),(2,6)\}$, then $A \subset P$.

Example A-2.13   If $S = \{(\theta,s)|\theta \text{ is any angle}, s \text{ is a real number}\}$ and $T = \{(30°,\frac{1}{2}),(45°,\frac{1}{2}), (60°,\frac{3}{2})\}$, then $T \subset S$.

A-3 Relations and Functions

A *relation* from set A to set B is any subset of $A \times B$. Thus a relation is a set of ordered pairs. A relation from A to A (usually we say a relation in A) is any subset of $A \times A$.

Example A-3.1   If $M = \{a,b,c\}$, $N = \{x,y,z\}$, and $B = \{(a,z),(c,x)\}$, then B is a relation from M to N.

Example A-3.2 If $Z = \{\text{integers}\}$, $N = \{\text{natural numbers}\}$, and $S = \{(-10,2),(-6,6),(2,4)\}$, then S is a relation from Z to N. $P = \{(2,-10),(2,4)\}$ is not a relation from Z to N. Why?

Example A-3.3 If $R = \{\text{real numbers}\}$ and $F = \{(x,y)|y = x^2\}$, then F is a relation in R since $F \subset R \times R$.

Exercise A-3.1

1 If $A = \{0,2,4\}$ and $B = \{5\}$, write a relation from A to B.

2 If $A = \{(x,y)|y = x + 2, x$ is any real number$\}$, complete the following ordered pairs defined by A:

   $(1,)$ $(-1,)$ $(\sqrt{3},)$ $(-\pi,)$ $(0,)$

3 If $A = \{\theta|\theta$ is any angle$\}$ and $R = \{\text{real numbers}\}$, write a relation from A to R.

4 If $C \subset A \times B$ and $C = \{(1,1),(2,4),(3,9),(4,16), \ldots \}$, describe A and B.

5 If $B = \{(x,y)|x^2 + y^2 = 9\}$, complete the following ordered pairs in B:

   $(0,)$ $(1,)$ $(,3)$ $(,-3)$ $(3,)$ $(,2)$

6 If $G = \{(4,2),(-4,2),(0,1),(1,0),(5,\frac{1}{2}),(5,\frac{1}{5})\}$, show why G is not a relation in the set of integers.

7 Tell which of the following defines a relation in R.

   (a) $\{(x,y)|x^2 = y - 4\}$

   (b) $\{(x,y)|y = |x^2 + 1|\}$

(c) $\left\{(r,t)\middle|r = \dfrac{t}{t-1},\, t \neq 1\right\}$

(d) $\{(\theta,a),(\phi,b)\}$

(e) $y = x^3$

(f) $H = \left\{\left(\dfrac{\pi}{6},\dfrac{1}{2}\right), \left(\dfrac{\pi}{4},\dfrac{\sqrt{2}}{2}\right), \left(\dfrac{\pi}{3},\dfrac{\sqrt{3}}{2}\right)\right\}$

The set of all first elements in the ordered pairs of a relation is called the *domain* of the relation. The set of all second elements in the ordered pairs of a relation is called the *range* of the relation.

Example A-3.4   If $M = \{a,b,c\}$, $N = \{x,y,z\}$, and $S = \{(a,z),(c,x)\}$, then S is a relation from M to N with domain $\{a,c\}$ and range $\{z,x\}$.

Example A-3.5   If $F = \{(x,y)|y = x^2\}$, then F is a relation in R, the domain of F is the set of real numbers, and the range of F is the set of nonnegative real numbers.

Exercise A-3.2   Give the domain and range of the following relations:

1 $N \times Z$

2 $Z \times N$

3 $\{(0,1),(0,2),(0,3),(0,4)\}$

4 $\{(1,2),(2,1)\}$

5 $\{0,1,2,3,4\} \times R$

6 $\{x|0 < x < 5, x \in N\} \times \{x|-1 < x < 1, x \in Z\}$

A *function* from set A to set B is a relation from A to B in which no two ordered pairs have the same first element. That is, each element in the domain is paired with a unique element in the range.

Example A-3.6 If $A = \{x,y,z\}$, $B = \{2,4,6, \ldots\}$, and $F = \{(x,2),(z,20)\}$, then F is a function from A to B.

Example A-3.7 If $G = \{(x,y)|x = y^2, y$ is a real number$\}$, then G is not a function in R, the set of real numbers, since $(4,2)$ and $(4,-2)$ belong to G.

Example A-3.8 If $H = \{(x,y)|x = y^4, y \in R\}$, then H is not a function in R since $(1,1)$ and $(1,-1)$ belong to H.

It is customary to represent functions such as $F = \{(x,y)|y = 2x + 1\}$ by other expressions. One could write equivalently $F = \{(x,f(x))|f(x) = 2x + 1\}$ or $f(x) = 2x + 1$. This notation is called "functional notation." The primary advantage of functional notation lies in the amount of information given when the second coordinate is expressed in the form $f(x)$. The f tells us which function we are working with, the x tells us what the first coordinate is, and $f(x)$ gives us the second coordinate. In $F = \{(x,f(x))|f(x) = 2x + 1\}$, the value of $f(x)$ for $x = 0$ is $f(0) = 2(0) + 1 = 1$, that is, $f(0) = 1$. This tells us that F is the function we are using and that if 0 is the first coordinate in an ordered pair belonging to F, its second coordinate is 1. Note that $f(2) = 5$, $f(K) = 2K + 1$, and $f(*) = 2* + 1$, provided that 5, K, and $*$ all belong to the domain of F. Indeed, if $f(x) = 2x + 1$, we may think of this as $F(\ ) = 2(\ ) + 1$, where any element in the domain of F may be placed in the parentheses.

Exercise A-3.3 **1** Show why $F = \{(x,y)|y = \pm\sqrt{x}\}$ is not a function in the set of real numbers.

2 For the relation F in problem 1, find y for the following values of x: 0, 4, 5, -12, a, b, π, $\sqrt{2}$, $a + b$, $x + h$.

3 Determine which of the following are functions in R.

(a) $\{(3,2),(2,5),(5,7),(7,9)\}$

(b) $\{(x,f(x))|f(x) = x\}$

(c) $\{(x,y)|y = \pm|x|\}$

(d) $\{(a,b),(c,d),(e,f),(a,g)\}$

(e) $\{(x,y)|x^2 + y^2 = 4\}$

(f) $\{(x,y)|x^2 - y + 2 = 5\}$

(g) $\{(x,y)|y^2 - x^2 - 4 = 0\}$

4 Can you give a rule of thumb for deciding which relations are functions?

5 In problem 3e above, find y for the following values of x: 0, 1, -1, $\sqrt{2}$, $r + s$.

6 If $x \in R$, does the set given in problem 3f define a function in the set of integers?

7 If $f(x) = 2x + 1$, find $\dfrac{f(x + h) - f(x)}{h}$.

If F is a function such that no two ordered pairs have the same second member, then the set of ordered pairs obtained by interchanging the elements in the ordered pairs of F also constitutes a function. It is called the *inverse function* of F, written F^{-1}.

Example A-3.9

$$F = \{(x,y),(z,w),(v,u)\}$$

$$F^{-1} = \{(y,x),(w,z),(u,v)\}$$

Example A-3.10

$$F = \{(x,y)|y = 2x\}$$

In F, $x = \dfrac{y}{2}$, and so interchanging the values x and y, we get

$$F^{-1} = \left\{(x,y)\Big|y = \frac{x}{2}\right\}$$

Example A-3.11

$$F = \{(1,2),(3,2),(4,5)\}$$

F^{-1} is not a function since $(1,2)$ and $(3,2)$ have the same second coordinate.

In example A-3.11 we see that $F^{-1} = \{(2,1),(2,3),(5,4)\}$ is not a function. It is, however, a mathematical relation. If we were to look at a subset of F, say $F_1 = \{(3,2),(4,5)\}$, F_1^{-1} is a function. By careful selection of a subset of the domain of F we are able to work with a function for which the inverse function exists. The concept of inverse functions plays a major role in the study of the trigonometric functions, and we may need to use our "trick" of restricting the domain of a function in order to have an inverse function exist.

The following steps may be used in finding the inverse of a function:

1 Switch variables, i.e., replace x with $f(x)$ and $f(x)$ with x in the given function.
2 Replace $f(x)$ with $f^{-1}(x)$.
3 Solve for $f^{-1}(x)$ in terms of x.

Example A-3.12

$$F = \{(x,f(x))|f(x) = x + 1\}$$

Switching variables makes $x = f(x) + 1$. Changing $f(x)$ to $f^{-1}(x)$ gives $x = f^{-1}(x) + 1$. Solving for $f^{-1}(x)$, we get $f^{-1}(x) = x - 1$. Hence,

$$F^{-1} = \{(x,f^{-1}(x))|f^{-1}(x) = x - 1\}$$

provided F^{-1} is a function.

Example A-3.13 $F = \{(x,f(x))|f(x) = x^2\}$

Switching variables, we get $x = [f(x)]^2$. Changing $f(x)$ to $f^{-1}(x)$ yields

$$x = [f^{-1}(x)]^2$$

Solving for $f^{-1}(x)$ gives $f^{-1}(x) = \pm\sqrt{x}$.

Exercise A-3.4   **1**   If $F = \{(a,b),(h,i),(r,s),(x,y)\}$, write F^{-1}.

2   If $F = \{(x,y)|y = 3x\}$, write F^{-1}.

3   If $G = \{(x,y)|y = 2x + 3\}$, write G^{-1}.

4   If $F = \{(x,f(x))|f(x) + 5x - 1 = 0\}$, write F^{-1}.

5   If $H = \{(x,h(x)|h(x) = 3^x\}$, find $h(0)$, $h(1)$, $h(-1)$, $h(2)$, $h^{-1}(9)$, $h^{-1}(3)$, and $h^{-1}(1)$.

6   If $F = \{(a,b),(b,b),(h,i),(b,i),(r,s)\}$, restrict the domain of F to form F_1, so that F_1^{-1} exists, and write F_1^{-1}.

7   If $F = \{(x,f(x))|[f(x)]^2 = 1 - x^2\}$, restrict the domain of F to form F_1, so that F_1^{-1} exists, and write F_1^{-1}.

A-4   Rectangular Coordinate System   The method of representing data pictorially or graphically is called *graphing*. The *graph of a relation* is the set of points which represent the ordered pairs in a relation. You probably recall that you can plot or represent these points in a rectangular coordinate (cartesian coordinate) system. This method of graphing functions and relations in the set R of real numbers employs a horizontal axis (x axis) and a vertical axis (y axis). Their intersection is the point which corresponds to the ordered pair $(0, 0)$, which is in $R \times R$.

Figure A-4.1

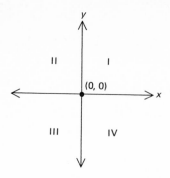

Figure A-4.1 shows that this intersection forms four quadrants in a plane where $x > 0$ and $y > 0$ in quadrant I; $x < 0$ and $y > 0$ in quadrant II; $x < 0$ and $y < 0$ in quadrant III; and $x > 0$ and $y < 0$ in quadrant IV. Hence, when we plot points which correspond to the ordered pairs of a relation in R, we are graphing a subset of $R \times R$.

Example A-4.1 Graph the relation $\{(-1,0),(2,3),(0,-4)\}$.

Solution

Figure A-4.2

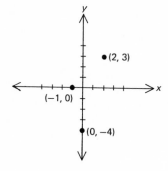

Example A-4.2 Graph $f = \{(x,y)|y = 2^x\}$.

Solution

x	0	1	-1	2	-2
y	1	2	$\frac{1}{2}$	4	$\frac{1}{4}$

Figure A-4.3

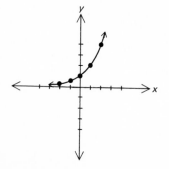

Exercise A-4.1   **1**   Graph $y = 2x - 7$.

2   Graph $2x - y = 5$.

3   Graph $y = x^2 + 5x + 6$.

4   Graph $y = (\frac{1}{2})^x$.

5   Graph $x^2 + y^2 = 16$.

6   Graph $y = 2x - \frac{3}{4}$.

Review Exercise   **1**   Use the roster method to describe the following:

(*a*)   The set of natural numbers greater than -1 and less than 5

(*b*)   The set of even integers greater than or equal to 0 and less than 10

(*c*)   The set of natural numbers whose squares are even integers less than 50

(*d*) $\{x|x \in N, 2x < 16\}$

(*e*) $\{x|x \in Z, -1 \leq x < 3\}$

2 Use set builder notation to describe the following:

(a) The set described in problem 1a

(b) The set described in problem 1b

(c) The set described in problem 1c

(d) The set of even multiples of 3 between 0 and 20

(e) The set of square roots of the even integers between 10 and 40

3 Write the cartesian product of the following:

(a) $A = \{1,2\}$, $B = \{a\}$

(b) $A = \{0,1\}$, $B = \{x,y,z\}$

(c) $A = \{w\}$, $B = \{t\}$

(d) $A = \{x \mid 2x < 12, x \in N\}$, $B = \{x \mid 1 \leq x < 3, x \in N\}$

(e) $A = \{x \mid -2 \leq 2x \leq 2, x \in Z\}$, $B = \{1, x, -1\}$

4 Which of the following are relations in the set of real numbers?

(a) $\{(1,1),(2,2),(3,3),(4,4)\}$

(b) $\{(1,3),(2,3),(3,3),(4,3)\}$

(c) $\{(3,1),(3,2),(3,3),(3,4)\}$

(d) $\{(x,y)|y = x^2\}$

(e) $\{(x,y)|y^2 = x\}$

5 Which of the following sets are functions?

(a) The set in problem 4*a*

(b) The set in problem 4*b*

(c) The set in problem 4*c*

(d) The set in problem 4*d*

(e) The set in problem 4*e*

6 Write the domain of the following relations:

(a) $\{(1,2),(3,1),(1,8),(2,7)\}$

(b) $\{(x,y),(a,b),(c,d)\}$

(c) $\{(x,y)|x \in N, y = 2x\}$

(d) $\{(x,y)|x \in N, 1 < x \le 5, y = 3x + 1\}$

(e) $\left\{(x,y)|x \in R, y = \dfrac{1}{\sqrt{x - 1}}\right\}$

(f) $\{(x,y)|x \in R, y = \sqrt{x - 1}\}$

(g) $\left\{(x,y)|x \in R, y = \dfrac{\sqrt{x}}{\sqrt{x - 4}}\right\}$

(h) $\{(x,y)|x \in N, y^2 = 4x\}$

(i) $\{(x,y)|x \in N, x^2 + y^2 = 4\}$

(j) $\left\{(x,y)|x \in Z, y = \dfrac{x}{x + 3}\right\}$

7   Write the range of the following:

(a)   The set described in problem 6b

(b)   The set described in problem 6d

(c)   The set described in problem 6f

(d)   The set described in problem 6e

(e)   The set described in problem 6h

8   Sketch the graphs of the following:

(a) $y = \frac{1}{2}x + 2$ (b) $y = 5$

(c) $x = 2$ (d) $y = x$

(e) $y = x^2 + 2x - 3$ (f) $y = |x + 1|$

(g) $y = 2^x$ (h) $y = \sqrt{9 - x^2}$

(i) $y = -\sqrt{9 - x^2}$ (j) $\{(x,y)|y = \pm\sqrt{x}, x \in N, 2 < x \leq 10\}$

Answers

Exercise A-2.1 **1** $\{2,4,6,8,10,12\}$ **2** $\{3,6,9, \ldots\}$
 3 $\{1,3,5,7,9,11,13,15\}$ **4** $\emptyset$
 5 $\{1,2,4\}$ **6** $\{5,6,3\}$

Exercise A-2.2 **1** $\{x|\frac{3}{11} < x < \sqrt{3}, x \in R\}$
 2 $P = \{x|f < x < m, x$ is a letter in the alphabet$\}$
 3 $\{(x,y)|y = \sqrt{2x^2 + 3}\}$
 4 $M = \{x|x$ is a positive odd integral multiple of 3$\}$
 5 $\{x|x$ is a positive integral multiple of 4$\}$
 6 $A = \{\theta|0° < \theta < 90°\}$

Exercise A-2.3 **1** $A \times B = \{(p,s),(p,t),(q,s),(q,t),(r,s),(r,t)\}; 6$
 2 $C = \{30°,60°\}$ $M = \{\frac{1}{2}, \sqrt{3}\}$
 $C \times M = \{(30°,\frac{1}{2}),(30°,\sqrt{3}),(60°,\frac{1}{2}),(60°,\sqrt{3})\}$
 $M \times C = \{(\frac{1}{2},30°),(\frac{1}{2},60°),(\sqrt{3},30°),(\sqrt{3},60°)\}$
 No, $C \times M$ is not equal to $M \times C$. There are four elements in each set.
 3 $R \times R = \{(x,y)|x, y \in R\}$
 4 If A contains n elements, $A \times A$ will contain $n \times n$, or n^2, elements.

Exercise A-3.1 **1** $\{(0,5),(2,5)\}$
 2 $A = \{(x,y)|y = x + 2, x$ is any real number$\}$
 $(1,3), (-1,1), (\sqrt{3},\sqrt{3} + 2), (-\pi,-\pi + 2), (0,2)$
 A is not a relation in the set of integers.
 A is a relation in the set of real numbers.
 3 $\{(30°,\frac{1}{2}),(59.6°,\sqrt{3}),(111.3°,\frac{3}{4})\}$
 4 $A = N, B = \{y|y = x^2, x \in N\}$
 5 $\{(0,3),(1,2\sqrt{2}),(0,3),(0,-3),(3,0),(\sqrt{5},2)\}$

6 $(5,\frac{1}{2}) \notin Z \times Z$

7 (*a*) Yes (*b*) Yes (*c*) Yes (*d*) No (*e*) Yes (*f*) Yes

Exercise A-3.2

1 Domain $= N$
Range $= Z$

2 Domain $= Z$
Range $= N$

3 Domain $= \{0\}$
Range $= \{1,2,3,4\}$

4 Domain $= \{1,2\}$
Range $= \{1,2\}$

5 Domain $= \{0,1,2,3,4\}$
Range $= R$

6 Domain $= 1,2,3,4$
Range $= 0$

Exercise A-3.3

1 $(1,1)$ and $(1,-1)$ both belong to F.

2 $f(0) = 0$
$f(4) = \pm 2$
$f(5) = \pm\sqrt{5}$
$f(-12)$ is not a member of R
$f(a) = \pm\sqrt{a}, a \geq 0$
$f(b) = \pm\sqrt{b}, b \geq 0$

$f(\pi) = \pm\sqrt{\pi}$
$f(\sqrt{2}) = \pm\sqrt{\sqrt{2}}$ or $\pm\sqrt[4]{2}$
$f(a + b) = \pm\sqrt{a + b}, a + b \geq 0$
$f(x + h) = \pm\sqrt{x + h}, x + h \geq 0$

3 (*a*) Yes (*b*) Yes (*c*) No (*d*) No (*e*) No (*f*) Yes (*g*) No

4 In order for a relation to be a function, for each value in the domain (x) there must be one and only one value in the range (y).

5 $f(0) = \pm 2$ $f(\sqrt{2}) = \pm\sqrt{2}$
$f(1) = \pm\sqrt{3}$ $f(r + s) = \pm\sqrt{4 - (r + s)^2}$
$f(-1) = \pm\sqrt{3}$

6 No

7 $\dfrac{2(x + h) + 1 - (2x + 1)}{h} = \dfrac{2x + 2h + 1 - 2x - 1}{h} = \dfrac{2h}{h} = 2$

Exercise A-3.4

1 $F^{-1} = \{(b,a),(i,h),(s,r),(y,x)\}$

2 $F^{-1} = \left\{(x,y)|y = \dfrac{x}{3}\right\}$

3 $G^{-1} = \left\{(x,g^{-1}(x))|g^{-1} = \dfrac{x - 3}{2}\right\}$

4 $F^{-1} = \left\{(x,f^{-1}(x))|f^{-1} = \dfrac{1 - x}{5}\right\}$

5 $h(0) = 1$ $h^{-1}(3) = 1$
$h(1) = 3$ $h^{-1}(1) = 0$
$h(-1) = \frac{1}{3}$
$h(2) = 9$
$h^{-1}(9) = 2$

6 $F_1^{-1} = \{(b,a),(i,h),(s,r)\}$

7 $F_1^{-1} = \{(x,f(x))|f(x) = \sqrt{1 - x^2}\}$

Exercise A-4.1

1

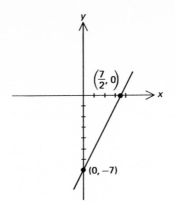

2

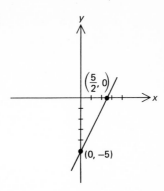

3

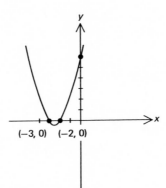

4

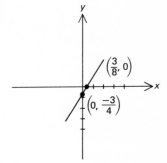

5

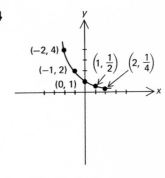

6

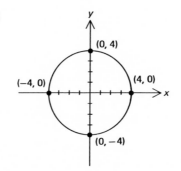

Review Exercise **1** (*a*) {1,2,3,4}
(*b*) {0,2,4,6,8}
(*c*) {2,4,6}
(*d*) {1,2,3,4,5,6,7}
(*e*) {−1,0,1,2}

2 (a) $\{x|-1 < x < 5, x \in N\}$
   (b) $\{x \in Z | x = 2n, n \in Z, 0 \le n < 5\}$
   (c) $\{x|x = 4n^2, n = 1, 2, 3\}$
   (d) $\{x|x = 6n, n = 1, 2, 3\}$
   (e) $\{x|x = \sqrt{2n}, n \in N, 5 < n < 20\}$

3 (a) $A \times B = \{(1,a),(2,a)\}$
   (b) $A \times B = \{(0,x),(0,y),(0,z),(1,x),(1,y),(1,z)\}$
   (c) $A \times B = \{(w,t)\}$
   (d) $A \times B = \{(1,1),(1,2),(2,1),(2,2),(3,1),(3,2),(4,1),(4,2),(5,1),(5,2)\}$
   (e) $A \times B = \{(-1,1),(-1,x),(-1,-1),(0,1),(0,x),(0,-1),(1,1),(1,x),(1,-1)\}$

4 All are relations in R.

5 (a) A function
   (b) A function
   (c) Not a function
   (d) A function
   (e) Not a function

6 (a) $\{1,3,2\}$
   (b) $\{x,a,c\}$
   (c) N
   (d) $\{x \in N | 1 < x \le 5\}$
   (e) All real numbers greater than 1
   (f) $\{x|x \in R, x \ge 1\}$
   (g) $\{x|x \in R, x > 4\}$
   (h) N
   (i) $\{1,2\}$
   (j) $\{x|x \in Z, x \ne -3\}$

7 (a) $\{y,b,d\}$
   (b) $\{7,10,13,16\}$
   (c) $\{y|y \in R, y \ge 0\}$
   (d) $\{y|y \in R, y > 0\}$
   (e) $\{y|y = \pm2\sqrt{x}, x \in N\}$

8 (a)

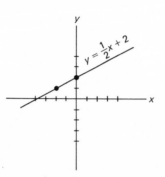

(b)

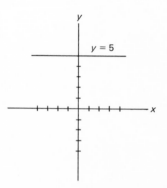

(c)

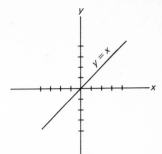

$x = 2$

(d)

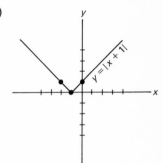

$y = x$

(e)

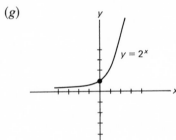

$y = x^2 + 2x - 3$

(f)

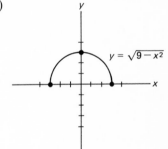

$y = |x + 1|$

(g)

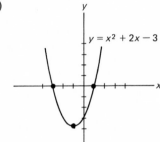

$y = 2^x$

(h)

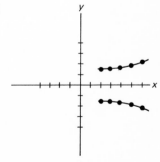

$y = \sqrt{9 - x^2}$

(i)

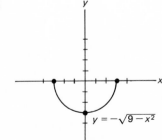

$y = -\sqrt{9 - x^2}$

(j)

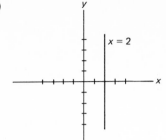

$\{(x, y) \mid y = \pm\sqrt{x}, x \in N, 2 < x \leqslant 10\}$

Appendix Logarithms

B

Objectives

Upon completion of Appendix B, you will be able to:

1. Change an exponential equation to logarithmic form. (B-2)

2. Change a logarithmic equation to exponential form. (B-2)

3. Given $\log_b N = L$, find N, L, or b in terms of the other numbers. (B-2)

4. Sketch the graph of $y = a^x$ for any positive number a. (B-3)

5. Sketch the graph of $y = \log_a x$ for any positive number a. (B-3)

6. Apply the properties of logarithms to evaluate logarithms of products, quotients, and roots. (B-4)

7. Given a common logarithm, identify the characteristic and mantissa. (B-5)

8. Find the common logarithm of a positive number. (B-6)

9. Find the antilog of a common logarithm. (B-7)

10. Use common logarithms to perform arithmetic computations. (B-8)

11. Change a logarithm from one base to another. (B-9)

12. Solve exponential equations. (B-11)

13. Solve logarithmic equations. (B-12)

B/Logarithms

B-1 Introduction

Logarithms have been in use for almost three hundred years. Today the use of logarithms for computation has been replaced for the most part by the electronic calculator. The importance of logarithms in higher mathematics remains fundamental. We shall study the properties of logarithms and use them in the solution of certain equations. In addition, we shall see how they are used in computation, and we shall note the relationship between logarithms and exponents.

B-2 Logarithms and Exponents

If we are given an equation of the form $y = b^x$, b is the base, x is the exponent, and y is the number that results when b is raised to the x power. In logarithmic terms, b is the base, y is the number, and x is the logarithm of y to the base b. Since $9 = 3^2$, 2 is the logarithm of 9 to the base 3.

If a positive number b different from 1 is raised to a power L to give the number N, the exponent L is called the *logarithm* of N and b is called the *base*. That is,

$$\log_b N = L$$

Notice that $\log_b N = L$ if and only if $b^L = N$.

Example B-2.1

$\log_2 8 = 3$ because $2^3 = 8$

Example B-2.2

$\log_3 9 = 2$ because $3^2 = 9$

Example B-2.3

If $3^a = y$, then $\log_3 y = a$.

Example B-2.4

If $b^2 = 7$, then $\log_b 7 = 2$.

If we are given an equation in logarithmic form, we must be able to change it to exponential form and vice versa. The following examples show problems of this nature.

Example B-2.5

Express $a^b = c$ in logarithmic form.

Solution

a is the base, c is the number, and b is the logarithm. Hence, $\log_a c = b$.

Example B-2.6   Express $\log_x 5 = 1$ in exponential form.

Solution x is the base, 5 is the number, and 1 is the exponent (log).   Hence, $x^1 = 5$.

Example B-2.7   Find N if $\log_5 N = 2$.

Solution   5 is the base and 2 is the exponent (log).   Hence, $5^2 = N = 25$.

.**Example B-2.8**   Find b if $\log_b 64 = 2$.

Solution   Change $\log_b 64 = 2$ to exponential form. b is the base, 2 is the exponent (log), and 64 is the number.   Hence,

$$b^2 = 64 \quad \text{and} \quad b = \pm\sqrt{64}$$

Thus $b = \pm 8$, but by definition $b > 0$ and so $b = 8$.

Example B-2.9   Find a if $\log_8 2 = a$.

Solution   Change $\log_8 2 = a$ to exponential form.   8 is the base, 2 is the number, and a is the exponent (log).   Hence, $8^a = 2$, but $8 = 2^3$.   So

$$(2^3)^a = 2^1$$
$$2^{3a} = 2^1 \quad \text{and} \quad a = \tfrac{1}{3}$$

Exercise B-2.1   Change each of the following to logarithmic form:

1 $2^3 = 8$

2 $x^2 = y$

3 $27^{1/3} = 3$

4 $16^{1/2} = 4$

5 $a^b = x$

Change each of the following to exponential form:

6 $\log_5 25 = 2$ **7** $\log_6 36 = 2$

8 $\log_b 5 = x$ **9** $\log_7 x = y$

10 $\log_3 n = p$

Find x if:

11 $3^x = 81$ **12** $32^x = 2$

13 $x^4 = 256$ **14** $x^3 = 125$

15 $\log_x 3 = \frac{1}{2}$ **16** $\log_x 32 = 5$

17 $\log_2 16 = x$ **18** $\log_5 25 = x$

19 $\log_{1/3} x = 7$ **20** $\log_5 x = \frac{1}{2}$

B-3 Graphs of $y = a^x$ **and** $y = \log_a x$

The graphs of $y = a^x$ and $y = \log_a x$ help illustrate the connection between exponents and logarithms. Notice that in the definition $\log_b N = L$, we said (with different symbols) that $y = \log_a x$ is the same as $a^y = x$. Now we switch the variables and write $a^x = y$. While $a^y = x$ is the same function as $y = \log_a x$, $a^x = y$ is the *inverse function* to $y = \log_a x$. This general form is graphed in Figure B-3.1. Notice the symmetry of the two curves with respect to the line $y = x$. This is characteristic of inverse functions.

Figure B-3.2 shows several exponential curves with different bases. Can you describe what effect changing the base has on the curve?

Figure B-3.3 shows several logarithmic curves with different bases. Can you describe what effect changing the base has on the curve?

Figure B-3.1

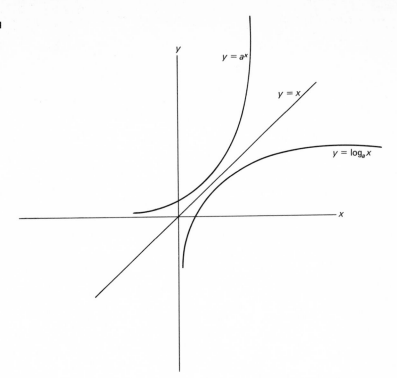

Figure B-3.2

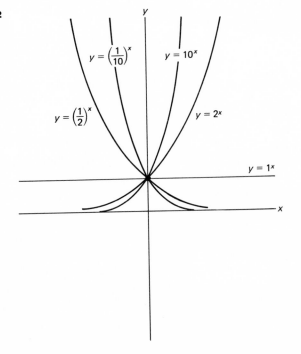

Figure B-3.3

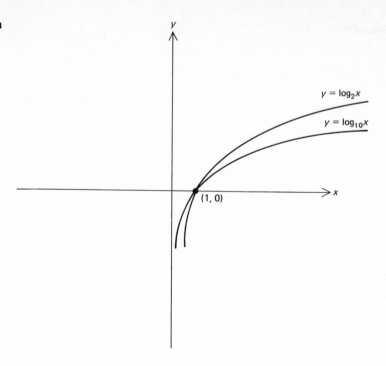

Exercise B-3.1 Sketch on the same coordinate system.

1 $y = 3^x$ and $y = \log_3 x$

2 $y = 5^x$ and $y = \log_5 x$

3 $y = 2^{3x}$ and $y = \frac{1}{3} \log_2 x$

4 $y = 10^x$ and $y = \log_{10} x$

5 $y = \frac{1}{2}^x$ and $y = \log_{1/2} x$

B-4 Properties of Logarithms

In this section we will study three basic laws of logarithms, which, along with our definition, will allow us to solve logarithmic and exponential equations and make arithmetic computations. In order to prove these properties (listed in Table B-4.1), we shall use the laws of exponents.

Table B-4.1

Numeric Form	Logarithmic Property
Product	$\log_b XY = \log_b X + \log_b Y$
Quotient	$\log_b \dfrac{X}{Y} = \log_b X - \log_b Y$
Exponent	$\log_b X^k = k \log_b X$

We now consider each property separately. The logarithm of the product of two numbers is equal to the sum of the logarithms of each factor.

$$\log_b XY = \log_b X + \log_b Y$$

Proof Let $\log_b X = x$ and $\log_b Y = y$. Applying the definition $\log_b X = x$ means $b^x = X$, and $\log_b Y = y$ means $b^y = Y$. Thus, $XY = b^x b^y = b^{x+y}$. Now, applying the definition once more, we find that $XY = b^{x+y}$ means

$$\log_b XY = x + y$$

Substitution gives $\log_b XY = \log_b X + \log_b Y$.

Example B-4.1 Find $\log_2 32$.

Solution
$$\log_2 8 = 3 \qquad \log_2 4 = 2$$
$$\log_2 32 = \log_2 (8 \cdot 4) = \log_2 8 + \log_2 4 = 3 + 2$$
$$\log_2 32 = 5$$

Check $2^5 = 32$

Example B-4.2 Find $\log_{10} 10$.

Solution If $\log_{10} 2 = 0.3010$ and $\log_{10} 5 = 0.6990$, then
$$\log_{10} 2 \cdot 5 = \log_{10} 2 + \log_{10} 5$$
$$= 0.3010 + 0.6990$$
$$= 1.0000$$

Check $10^1 = 10$

The logarithm of the quotient of two numbers is equal to the logarithm of the dividend minus the logarithm of the divisor.

$$\log_b \frac{X}{Y} = \log_b X - \log_b Y$$

Proof Let $\log_b X = x$ and $\log_b Y = y$. Applying the definition, $\log_b X = x$

means $X = b^x$, and $\log_b Y = y$ means $Y = b^y$. Thus, $\dfrac{X}{Y} = \dfrac{b^x}{b^y} = b^{x-y}$.

Now, applying the definition once more, $\dfrac{X}{Y} = b^{x-y}$ means $\log_b \dfrac{X}{Y} = x - y$.

Substitution gives

$$\log_b \frac{X}{Y} = \log_b X - \log_b Y$$

Example B-4.3   Find $\log_2 \frac{16}{4}$.

Solution $\log_2 16 = 4 \qquad \log_2 4 = 2$

$\log_2 \frac{16}{4} = \log_2 16 - \log_2 4$

$= 4 - 2$

$= 2$

Check $2^2 = 4 = \frac{16}{4}$

Example B-4.4   Find $\log_{10} \frac{10}{2}$.

Solution   If $\log_{10} 2 = 0.3010$ and $\log_{10} 10 = 1$,

$\log_{10} \frac{10}{2} = \log_{10} 10 - \log_{10} 2$

$= 1 - 0.3010$

$= 0.6990$

Check   From Example B-4.2, $\log_{10} 5 = 0.6990$.

The logarithm of a power of a number is equal to the product of the exponent and the logarithm of the number.

$$\log_b X^k = k \log_b X$$

Proof   Let $\log_b X = x$. Applying our definition of logarithms, we find that

$\log_b X = x$ means $X = b^x$

Raising each member of $X = b^x$ to the kth power, we get

$X^k = (b^x)^k$

$= b^{xk}$

$= b^{kx}$

Applying the definition, we see that $X^k = b^{kx}$ means $\log_b X^k = kx$. Substitution gives $\log_b X^k = k \log_b X$.

Example B-4.5   Find $\log_2 4^3$.

Solution

$$\log_2 4 = 2$$

$$\log_2 4^3 = 3 \log_2 4$$

$$= 3 \cdot 2$$

$$= 6$$

Check

$$\log_2 4^3 = \log_2 64$$

$$= \log_2 2^6$$

$$= 6$$

Example B-4.6   Find $\log_{10} 2^3$.

Solution

$$\log_{10} 2^3 = 3 \log_{10} 2$$

$$= 3(0.3010)$$

$$= 0.9010$$

Example B-4.7   Find $y = \log_3 \sqrt{\dfrac{27 \times 243}{81}}$.

Solution

$$y = \log_3 \left(\frac{27 \times 243}{81}\right)^{1/2}$$

$$= \tfrac{1}{2}\log_3 \left(\frac{27 \times 243}{81}\right)$$

$$= \tfrac{1}{2}[(\log_3 27 + \log_3 243) - \log_3 81]$$

$$= \tfrac{1}{2}[(3 + 5) - 4]$$

$$= \tfrac{1}{2}(8 - 4)$$

$$= 2$$

Exercise   B-4.1   Use $\log_3 9 = 2$, $\log_3 27 = 3$, $\log_3 81 = 4$, $\log_3 243 = 5$, $\log_3 729 = 6$, $\log_3 2187 = 7$, and the properties of logarithms to find the value of y:

1 $y = \log_3 (9 \times 729)$

2 $y = \log_3 (27 \times 81 \times 2187)$

3 $y = \log_3 (243 \times 9 \times 729 \times 27)$

4 $y = \log_3 (9 \times 9 \times 27 \times 729 \times 729)$

5 $y = \log_3 \frac{81}{9}$

6 $y = \log_3 \left(\frac{81 \times 729}{243} \right)$

7 $y = \log_3 \left(\frac{2187 \times 9}{27} \right)$

8 $y = \log_3 \left(\frac{729}{9 \times 81} \right)$

9 $y = \log_3 27^2$

10 $y = \log_3 9^5$

11 $y = \log_3 (2187)^5$

12 $y = \log_3 \sqrt{729}$ **Hint:** $\sqrt[n]{X^m} = X^{m/n}$

13 $y = \log_3 \sqrt[5]{243}$

14 $y = \log_3 \sqrt[3]{729}$

15 $y = \log_3 \sqrt{\dfrac{729 \times 81}{9}}$

B-5 Common or Briggs Logarithms

When logarithms are used in arithmetic computations, any positive number other than 1 may be used for the base. In practice, the number 10 and the irrational number e ($e = 2.718 \ldots$) are usually used as bases. When 10 is the base, the logarithmic system is called the "common" or "Briggs" system. When e is used, the system is called the "natural" or "Napierian" system.

We now examine the common, or Briggs, system of logarithms. Consider

$$\log_{10} 212 = 2.3263$$

$\log_{10} 212 = 2.3263$ means $212 = 10^{2.3263}$ by definition. We write

$$\log_{10} 212 = 2.3263$$

When 10 is the base, it is omitted from the notation. Notice that 2.3263 is actually $2 + 0.3263$. 2 is called the "characteristic," and 0.3263 is called the "mantissa."

Observe that $\log_{10} 1.4 = 0.1461$ means $1.4 = 10^{0.1461}$ by definition. We write $\log_{10} 1.4 = 0.1461$, where 0 is the characteristic and 0.1461 is the mantissa. Additionally, $\log_{10} 31.1 = 1.4928$ means $31.1 = 10^{1.4928}$ by definition. We write $\log 31.1 = 1.4928$, where 1 is the characteristic and 0.4928 is the mantissa.

We now define the characteristic and mantissa of a common logarithm.

If L is the common logarithm of a number and $L = N + n$, where N is an integer and n is a decimal fraction, $0 \le n < 1$, then N is the characteristic of L and n is the mantissa.

B-6 Finding the Common Logarithm of a Given Number

Every positive number may be expressed as the sum of a number between 1 and 10 and a power of 10. A number expressed in this form is said to be in "scientific notation."

The following examples illustrate scientific notation:

Example B-6.1

Express 341 in scientific notation.

Solution

$$341 = 3.41 \times 10^2$$

Example B-6.2 Express 0.00341 in scientific notation.

Solution $0.00341 = 3.41 \times 10^{-3}$

We now use scientific notation to find the common logarithms of given numbers. Each given number is written as the sum of a number between 1 and 10 and a power of 10. The common logarithm of the given number is the sum of the power to which 10 is raised (characteristic) and the logarithm of the number between 1 and 10 (mantissa). The logarithms of all three-digit numbers between 1 and 10 are given in Table C-3; hence Table C-3 is a table of mantissas.

The following examples illustrate the techniques of finding the common logarithms of a given number.

Example B-6.3 Find log 244.

Solution
$$244 = 2.44 \times 10^2$$

$$\log 244 = \log 2.44 + \log 10^2$$

$$= 0.3874 \text{ (mantissa)} + 2 \text{ (characteristic)}$$

$$= 2.3874$$

To find the mantissa (log 2.44), we located the entry in Table C-3 in the row headed 24 and under the column headed 4.

Example B-6.4 Find log 0.00393.

Solution
$$0.00393 = 3.93 \times 10^{-3}$$

$$\log 0.00393 = \log 3.93 \text{ (mantissa)} + \log 10^{-3} \text{ (characteristic)}$$

$$= 0.5944 + (-3)$$

Since 0.5944 is positive and $0.5944 + (-3) \neq -3.5944$, we express -3 as $7 - 10$ and write log $0.00393 = 7.5944 - 10$.

Example B-6.5 Find log 8466.

Solution
$$8466 = 8.466 \times 10^3$$

$$\log 8466 = \log 8.466 \text{ (mantissa)} + \log 10^3 \text{ (characteristic)}$$

$$= 0.9277 + 3 = 3.9277$$

Since Table C-3 only gives the logarithms of three-digit numbers between 1 and 10, in the previous example we had to use linear interpolation to find log 8.466. The method of interpolation to find log 8.466 is given below.

$$0.010 \begin{bmatrix} \begin{array}{c} \text{log } 8.470 = 0.9279 \\ 0.006 \begin{bmatrix} \text{log } 8.466 = x \\ \text{log } 8.460 = 0.9274 \end{bmatrix} d \end{array} \end{bmatrix} 0.0005$$

$$\frac{0.006}{0.010} = \frac{d}{0.0005}$$

$$0.0003 = d$$

$$x = 0.9274 + d$$

$$= 0.9274 + 0.0003$$

$$\text{log } 8.466 = 0.9277$$

Example B-6.6 Find log 0.04837.

Solution $0.04837 = 4.837 \times 10^{-2}$

$\text{log } 0.04837 = \text{log } 4.837 \text{ (mantissa)} + \text{log } 10^{-2} \text{ (characteristic)}$

To find log 4.837, we must interpolate as follows:

$$0.010 \begin{bmatrix} \begin{array}{c} \text{log } 4.840 = 0.6849 \\ 0.007 \begin{bmatrix} \text{log } 4.837 = x \\ \text{log } 4.830 = 0.6839 \end{bmatrix} d \end{array} \end{bmatrix} 0.0010$$

$$\frac{0.007}{0.010} = \frac{d}{0.0010}$$

$$d = 0.0007$$

$$x = 0.6839 + d$$

$$= 0.6839 + 0.0007$$

$$= 0.6846$$

Hence

$$\text{log } 0.04837 = 0.6846 + (-2)$$

$$= 8.6846 - 10$$

Exercise   B-6.1   Find the characteristics of the common logarithms of the following numbers:

1 275   **2** 8

3 14   **4** 57.1

5 1200   **6** 0.42

7 0.775   **8** 0.007

9 0.0004   **10** 0.0054

Find the mantissas of the common logarithms of the following numbers:

11 2.32   **12** 4.11

13 2390   **14** 575

15 5.21   **16** 25.75

17 14.37   **18** 30.05

19 271.2   **20** 2173

Find the common logarithms of the following numbers:

21 2.47   **22** 9.89   **23** 0.0054

24 0.0875   **25** 0.000983   **26** 0.004151

27 8216   **28** 14.75   **29** 1.844

30 7.605

B-7 Finding the Antilog of a Given Number

Before we can use logarithms for arithmetic computation, we must be able to find a number whose logarithm is given. This is called "finding the antilog."

If $\log N = L$, then N is the antilog of L.

To find the antilog of a given number, we locate the given mantissa in Table C-3 and find the number between 1 and 10 for which it is the logarithm. Interpolation may be necessary. We then change the position of the decimal to correspond to the characteristic of the given logarithm.

The following examples illustrate the procedure.

Example B-7.1   If $\log N = 2.5866$, find N.

Solution   From Table C-3

$$\log 3.86 = 0.5866$$

Since $\log N$ has characteristic 2,

$$\log N = 0.5866 + 2$$
$$N = 3.86 \times 10^2$$
$$N = 386$$

Example B-7.2   If $\log N = 8.9270 - 10$, find N.

Solution   From Table C-3 and interpolation, we find

$$\log 8.452 = 0.9270$$
$$\log N = 0.9270 + (-2)$$
$$N = 8.452 \times 10^{-2}$$
$$N = 0.08452$$

Exercise B-7.1   Find N if $\log N$ is:

1   2.5172

2   8.3692 − 10

3   0.5105

4   1.5855

5   9.9872 − 10

6   0.9515

7 2.9426   **8** 6.5424

9 $7.7220 - 10$ **10** $9.2390 - 10$

B-8 Computation with Common Logarithms

When an arithmetic computation is to be performed with logarithms, the logarithm of each number is found, and by using the properties of logarithms these are combined into a single logarithmic value. The result of the computation is the antilog of this logarithmic value.

The following examples illustrate the procedure.

Example B-8.1
Computation

Compute $N = 40 \times 118$.

$$\log 40 = 1.6021$$

$$\log 118 = 2.0719$$

$$\log N = \log 40 + \log 118$$

$$= 1.6021 + 2.0719$$

$$= 3.6740$$

$$N = \text{antilog } 3.6740$$

$$\log N = 0.6740 + 3$$

$$= \log 4.72 + \log 10^3$$

$$N = 4.72 \times 10^3$$

$$N = 4720$$

Example B-8.2
Computation

Compute $N = \frac{1284}{321}$.

$$\log 1284 = 3.1086 \qquad \text{(by interpolation)}$$

$$\log 321 = 2.5065$$

$$\log N = \log 1284 - \log 321$$

$$= 3.1086 - 2.5065$$

$$= 0.6021$$

$$N = \text{antilog } 0.6021$$

$$N = 4$$

Example B-8.3   Compute $N = \sqrt[3]{248}$.

Computation

$$\log 248 = 2.3945$$

$$N = \sqrt[3]{248}$$

$$= (248)^{1/3}$$

$$\log N = \log (248)^{1/3}$$

$$= \tfrac{1}{3} \log 248$$

$$= \tfrac{1}{3}(2.3945)$$

$$= 0.7982$$

$$N = \text{antilog } 0.7982$$

$$= 6.283 \qquad \text{(by interpolation)}$$

Example B-8.4   Compute $\sqrt[4]{\dfrac{5312 \times \sqrt{2.689}}{(61.28)^2 \times 41.67}}$.

Computation

$$\log 5312 = 3.7252$$

$$\log 2.689 = 0.4296$$

$$\log 61.28 = 1.7874$$

$$\log 41.67 = 1.6198$$

$$N = \left[\frac{5312 \times (2.689)^{1/2}}{(61.28)^2 \times 41.67} \right]^{1/4}$$

$$\log N = \tfrac{1}{4}\{[\log 5312 + \tfrac{1}{2}(\log 2.689)] - (2\log 61.28 + \log 41.67)\}$$

$$= \tfrac{1}{4}\{[3.7252 + \tfrac{1}{2}(0.4296)] - [2(1.7874 + 1.6198)]\}$$

$$= \tfrac{1}{4}[(3.7252 + 0.2148) - (3.5748 + 1.6198)]$$

$$= \tfrac{1}{4}(3.9400 - 5.1946)$$

Now since 5.1946 is greater than 3.9400 and we won't be able to divide by 4, we add and then subtract 40.

$$\log N = \tfrac{1}{4}(43.9400 - 5.1946 - 40)$$

$$= \tfrac{1}{4}(38.7454 - 40)$$

$$= 9.6863 - 10$$

$$N = \text{antilog } 9.6863 - 10$$

$$N = 0.4856$$

Exercise B-8.1   Use logarithms to compute N in the following problems:

1 $N = 421 \times 375$ **2** $N = 0.0475 \times 842$

3 $N = 0.0038 \times 0.0042$ **4** $N = \frac{720}{24}$

5 $N = \frac{3400}{172}$ **6** $N = \frac{0.00573}{0.0241}$

7 $N = \frac{2.95 \times 59.9}{428}$ **8** $N = \frac{76.2 \times 843}{30.9}$

9 $N = 4.44^4$ **10** $N = \sqrt[4]{1223}$

11 $N = 2.07^2$ **12** $N = \sqrt[3]{1470}$

13 $N = \sqrt{\sqrt[4]{348}}$ **14** $N = \frac{\sqrt{845}}{\sqrt{0.0742}}$

15 $N = 42.1\sqrt[5]{0.0321}$ **16** $N = 59.03\sqrt[4]{0.03579}$

17 $N = \left(\frac{2.05}{1.11}\right)^{4/3}$ **18** $N = \sqrt[3]{\frac{(540)^2\sqrt{241}}{(321)^4}}$

19 $N = \sqrt{\frac{\sqrt{28.44}\sqrt{0.5743}}{453.6}}$ **20** $N = \sqrt{\frac{\sqrt{44.81}(0.5421)^2}{\sqrt{(4142)^3(0.4118)^3}}}$

B-9 Changing Bases   The base of a logarithmic system should be chosen so as to simplify computation. The use of base 10 is convenient for arithmetic computations, and the use of base e is convenient for much scientific work. Since various bases may be encountered, we must be able to change from one base to another.

The logarithm of a number N to a base b is equal to the quotient of the logarithm of the number N to a new base c divided by the logarithm of the old base b to the new base c.

$$\log_b N = \frac{\log_c N}{\log_c b}$$

Example B-9.1 Express $\log_5 4$ in terms of logarithms to the base 3.

Solution $$\log_5 4 = \frac{\log_3 4}{\log_3 5}$$

Example B-9.2 Use common logarithms to find $\log_e 45$.

Solution $$\log_e 45 = \frac{\log 45}{\log e}$$

$$= \frac{1.6532}{0.4343}$$

$$\log (\log_e 45) = \log 1.6532 - \log 0.4343$$

$$= 0.2183 - (0.6378 - 1)$$

$$= 1.2183 - 0.6378$$

$$= 0.5805$$

$$\log_e 45 = \text{antilog } 0.5805$$

$$= 3.8067$$

Example B-9.3 Find $\log_7 326$.

Solution $$\log_7 326 = \frac{\log 327}{\log 7}$$

$$= \frac{2.5145}{0.8451}$$

$$= 2.975$$

Exercise B-9.1   Find N.

1 $N = \log_3 17$ **2** $N = \log_e 347$

3 $N = \log_7 e$ **4** $N = \log_e 71$

5 $N = \log_{142} 3147$ **6** $N = \log_3 6.83$

7 $N = \log_5 9.23$ **8** $N = \log_8 1077$

9 $N = \log_e (34.14)^2$ **10** $N = \log_{17} 0.0075$

B-10 Computation with Natural Logarithms

Computations with natural logarithms can be performed directly by using a table of natural logarithms. However, when calculations with natural logarithms are required, the task can be accomplished by simply changing the base to 10 and using a table of common logarithms.

Although natural logarithms do not have the convenience of "characteristics" and "mantissas," the procedure for finding the natural logarithm of a number is very similar to the procedure used in the system of common logarithms.

In order to find the natural logarithm of a number N, we first write the number in scientific notation:

$$N = n \times 10^k \qquad 1 \le n < 10, \, k \in Z$$

where Z is the set of integers. Hence, the natural logarithm of N is the sum of the natural logarithm of n and the natural logarithm of 10^k.

If $N = n \times 10^k$, $1 \le n < 10$, $k \in Z$, then

$$\log_e N = \log_e n + \log_e 10^k$$

$$= \log_e n + k \log_e 10$$

$$= \log_e n + 2.3026k$$

It is common to write $\log_e x$ as $\ln x$.

B-11 Exponential Equations

An exponential equation is an equation in which a variable occurs in one or more exponents. The following are examples of exponential equations:

$$2^x = 8$$

$$3^{(x+2)} = 27$$

$$2^{(x+5)} = 5^{(x+2)}$$

Although common or natural logarithms are widely used, we may choose any base that would simplify the work of solving a given equation.

The following examples illustrate the method of solving exponential equations.

Example B-11.1 Solve $2^x = 64$.

Solution Since $64 = 2^6$, we choose 2 for our logarithmic base.

$$2^x = 64 \qquad \log_2 2^x = \log_2 2^6$$

$$2^x = 2^6 \qquad x \log_2 2 = 6 \log_2 2$$

$$x \cdot 1 = 6 \cdot 1 \qquad SS = \{6\}$$

Example B-11.2 Solve $3^x = 125$.

Solution Since 125 is not a power of 3, we choose to use common logarithms.

$$3^x = 125$$

$$\log 3^x = \log 125$$

$$x \log 3 = \log 125$$

$$x = \frac{\log 125}{\log 3}$$

$$x = \frac{2.0969}{0.4771}$$

$$x = 4.40$$

$$SS = \{4.40\}$$

Caution Through the use of logarithms, $x = \dfrac{2.0969}{0.4771}$ may be solved but $x \neq 2.0969 - 0.4771$.

Example B-11.3 Solve $4^{x+4} = 3^{x+2}$.

Solution Using common logs:

$$4^{x+4} = 3^{x+2}$$

$$\log 4^{x+4} = \log 3^{x+2}$$

$$(x + 4) \log 4 = (x + 2) \log 3$$

$$x \log 4 + 4 \log 4 = x \log 3 + 2 \log 3$$

$$x \log 4 - x \log 3 = 2 \log 3 - 4 \log 4$$

$$x(\log 4 - \log 3) = 2 \log 3 - 4 \log 4$$

$$x = \frac{2 \log 3 - 4 \log 4}{\log 4 - \log 3}$$

$$x = \frac{2(0.4771) - 4(0.6021)}{0.6021 - 0.4771}$$

$$x = \frac{0.9542 - 2.4084}{0.1250}$$

$$= \frac{-1.4542}{0.1250}$$

$$= -11.6$$

$$SS = \{-11.6\}$$

Exercise B-11.1 Use logarithms to solve the following equations:

1 $5^x = 625$ **2** $7^{2x} = 49$

3 $2^{3x+1} = 128$ **4** $3^{5x-1} = 81$

5 $8^{x^2-x} = 64$ **6** $5^x = 185$

7 $75^x = 3.47$ **8** $52.6^x = 6.85$

9 $173^x = 248$ **10** $5.32^x = 344$

B-12 Solving Logarithmic Equations

In general, the method used for solving equations that contain logarithms to a common base has the following steps:

1. Apply the properties of logarithms to simplify the equation into a single logarithm of the form $\log_a N = L$.

2. Apply the definition of the logarithm to get the exponential form

$$a^L = N$$

3. Solve $a^L = N$, being careful to exclude any potential solutions that would make N negative.

The following examples illustrate the procedure.

Example B-12.1 Solve $\log_5 (x - 1) + \log_5 (x + 3) = 1$.

Solution Applying properties of logarithms, we find

$$\log_5 (x - 1) + \log_5 (x + 3) = 1$$

$$\log_5 (x - 1)(x + 3) = 1$$

$$\log_5 (x^2 + 2x - 3) = 1$$

Since $\log_b N = L$ means $b^L = N$,

$$x^2 + 2x - 3 = 5$$

$$x^2 + 2x - 8 = 0$$

$$(x + 4)(x - 2) = 0$$

$$x = -4 \quad \text{or} \quad x = 2$$

However, if $x = -4$, then $x - 1 < 0$. Hence $\log_5 (x - 1)$ is undefined and -4 must not be included in our solution set. Thus SS = {2}.

Example B-12.2 Solve $\log_6 3 + \log_6 (x + 6) = 2$.

Solution

$$\log_6 3 + \log_6 (x + 6) = 2$$

$$\log_6 3(x + 6) = 2$$

$$3(x + 6) = 6^2$$

$$3x + 18 = 36$$

$$3x = 18$$

$$x = 6$$

$$\text{SS} = \{6\}$$

Example B-12.3 Solve $\log_5 (3x + 7) - \log_5 (x - 5) = 2$.

Solution $\log_5 (3x + 7) - \log_5 (x - 5) = 2$

$$\log_5 \frac{3x + 7}{x - 5} = 2$$

$$\frac{3x + 7}{x - 5} = 5^2$$

$$3x + 7 = 25(x - 5)$$

$$3x + 7 = 25x - 125$$

$$132 = 22x$$

$$6 = x$$

$$SS = \{6\}$$

Exercise B-12.1 Find the solution set in each problem.

1 $\log_5 2 + \log_5 (x + 1) = 3$

2 $\log_2 3 + \log_2 (2x - 1) = 8$

3 $\log_3 (2x + 1) - \log_3 (x - 1) = 3$

4 $\log_8 (x + 1) + \log_8 (x - 1) = 2$

5 $\log_4 (3 - x) + \log_4 2x = 1$

6 $\log_2 (3x + 1) - \log_2 2x = 4$

7 $\log_2 5 + \log_2 2x = 6$

Review Exercise **1** Express in logarithmic form:

(a) $y = 6^x$

(b) $y = 4^2$

(c) $64 = 2^x$

(d) $k = a^b$

2 Express in exponential form:

(a) $y = \log_3 90$

(b) $y = \log_a 72$

(c)　$y = \log_{10} 42$　　　　　　　　(d)　$y = \log_7 99$

3　Solve for x:

(a)　$\log_5 x = 2$　　　　　　　　(b)　$\log_7 49 = x$

(c)　$\log_6 x = 3$　　　　　　　　(d)　$\log_3 81 = x$

(e)　$\log_x 49 = 2$　　　　　　　　(f)　$\log_x 2048 = 11$

4　Sketch the graph of:

(a)　$y = 3^x$　　　　　　　　(b)　$y = (\tfrac{1}{4})^x$

(c)　$y = 6^{2x}$　　　　　　　　(d)　$y = (\tfrac{1}{10})^x$

5　Sketch the graph of:

(a)　$y = \log_2 x$　　　　　　　　(b)　$y = \log_7 x$

(c)　$y = \log_e x$　　　　　　　　(d)　$y = \log_{1/2} x$

6　Given $\log_5 25 = 2$, $\log_5 3125 = 5$ and $\log_5 15625 = 6$, compute:

(a)　$y = \log_5 (25)(3125)$　　　　　(b)　$y = \log_5 \left[\dfrac{(3125)(15,625)}{25} \right]$

(c)　$y = \log_5 (25)^7$　　　　　　　(d)　$y = \log_5 \left[\dfrac{(15,625)^3}{25} \right]$

7　Give the characteristics and mantissa of the following:

(a)　$\log x = 2.1414$　　　　　　　(b)　$\log x = 11.1345$

(c)　$\log x = 8.7124 - 10$　　　　　(d)　$\log x = 5.8814 - 10$

8 Use Table C-3 to find:

(a) log 714

(b) log 0.0752

(c) log 1.34

(d) log 21.75

(e) log 0.03147

(f) log 0.0003147

9 Use Table C-3 to find:

(a) $y = $ antilog 2.9731

(b) $y = $ antilog $9.9881 - 10$

(c) $y = $ antilog $7.9300 - 10$

(d) $y = $ antilog 3.6880

10 Use Table C-3 to compute x:

(a) $x = (411)(721)$

(b) $x = (0.342)(0.0721)$

(c) $x = \dfrac{451}{71}$

(d) $x = \dfrac{0.0215}{0.00511}$

(e) $x = \sqrt{714}$

(f) $x = \sqrt[3]{0.0775}$

(g) $x = \dfrac{(314)(0.075)}{42}$

(h) $x = \sqrt[3]{\dfrac{(741)(2314)}{0.0777}}$

11 Express the following in base e:

(a) $\log_2 14$

(b) $\log 88$

(c) $\log_{10} 75$

(d) $\log_x y$

(e) $\log_5 a$

(f) $\log_4 57$

12 Solve the following:

(a) $5^x = 625$

(b) $8^x = 512$

(c) $6^x = 121$

(d) $7^x = 3147$

13 Solve the following:

(a) $\log_5 (x - 24) + \log_5 x = 2$

(b) $\log_3 (x + 6) + \log_3 x = 3$

(c) $\log_6 (2x + 1) - \log_6 (x + 2) = 1$

(d) $\log_7 (x - 1) - \log_7 x = 1$

Answers

Exercise B-2.1

1 $\log_2 8 = 3$

3 $\log_{27} 3 = \frac{1}{3}$

5 $\log_a x = b$

7 $6^2 = 36$

9 $7^y = x$

11 4

13 4

15 9

17 4

19 $(\frac{1}{3})^7$

2 $\log_x y = 2$

4 $\log_{16} 4 = \frac{1}{2}$

6 $5^2 = 25$

8 $b^x = 5$

10 $3^p = n$

12 $\frac{1}{5}$

14 5

16 2

18 2

20 $\sqrt{5}$

Exercise B-3.1 **1**

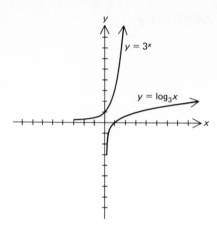

2

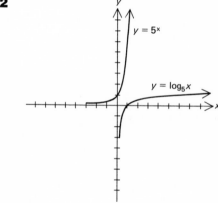

3

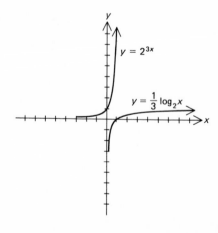

4

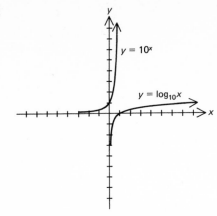

5

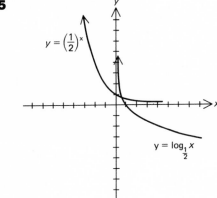

Exercise B-4.1	**1** 8	**2** 14
	3 16	**4** 19
	5 2	**6** 5
	7 6	**8** 0
	9 6	**10** 10
	11 35	**12** 3
	13 1	**14** 2
	15 4	

Exercise B-6.1	**1** 2	**2** 0
	3 1	**4** 1
	5 3	**6** −1

7	−1		**8**	−3
9	−4		**10**	−3
11	0.3655		**12**	0.6138
13	0.3784		**14**	0.7597
15	0.7168		**16**	0.4107
17	0.1576		**18**	0.4779
19	0.4333		**20**	0.3371
21	0.3927		**22**	0.9952
23	7.7324 − 10		**24**	8.9420 − 10
25	6.9926 − 10		**26**	7.6181 − 10
27	3.9146		**28**	1.1688
29	0.2658		**30**	0.8811

Exercise B-7.1	**1**	329	**2**	0.0234
	3	3.24	**4**	38.5
	5	0.971	**6**	8.944
	7	876.2	**8**	3486000
	9	0.005272	**10**	0.1736

Exercise B-8.1	**1**	157,900	**2**	40.00
	3	0.00001596	**4**	30
	5	19.77	**6**	0.2378
	7	0.4123	**8**	2079
	9	388.7	**10**	5.912
	11	4.285	**12**	11.37
	13	2.104	**14**	106.7
	15	21.16	**16**	25.67
	17	2.266	**18**	0.0753
	19	0.0944	**20**	0.005286

Exercise B-9.1	**1**	2.5789	**2**	5.835
	3	0.5139	**4**	4.2627
	5	1.6252	**6**	1.7489
	7	1.381	**8**	3.3574
	9	7.0609	**10**	−1.727

Exercise B-11.1	**1**	SS = {4}	**2**	SS = {1}
	3	SS = {2}	**4**	SS = {1}

5 SS = {−1,2}   **6** SS = {3.243}
7 SS = {0.2881}   **8** SS = {0.4856}
9 SS = {1.0699}   **10** SS = {3.498}

Exercise B-12.1
1 SS = {61.5}   **2** SS = {43.17}
3 SS = {$\frac{28}{25}$}   **4** SS = {$\sqrt{65}$}
5 SS = {1}   **6** SS = {$\frac{1}{29}$}
7 SS = {6.4}

Review Exercise
1 (*a*) $\log_6 y = x$ (*b*) $\log_4 y = 2$
    (*c*) $\log_2 64 = x$ (*d*) $\log_a k = b$
2 (*a*) $3^y = 90$ (*b*) $a^y = 72$
    (*c*) $10^y = 42$ (*d*) $7^y = 99$
3 (*a*) 25   (*b*) 2
    (*c*) 216   (*d*) 4
    (*e*) 7   (*f*) 2
4 (*a*)

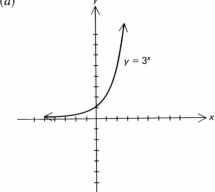

$y = 3^x$

(*b*)

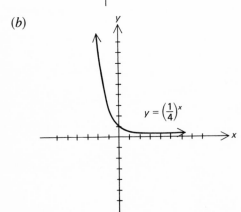

$y = \left(\frac{1}{4}\right)^x$

(c)

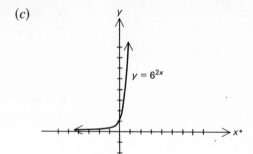

$y = 6^{2x}$

(d)

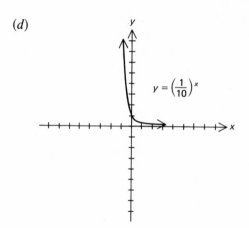

$y = \left(\dfrac{1}{10}\right)^x$

5 (a)

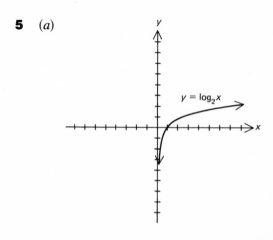

$y = \log_2 x$

(b)

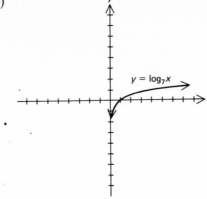

$y = \log_7 x$

(c)

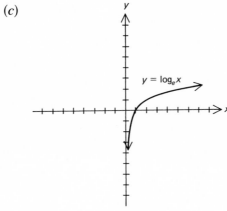

$y = \log_e x$

(d)

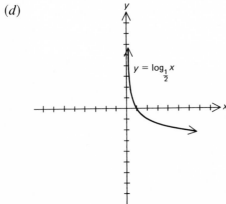

$y = \log_{\frac{1}{2}} x$

6 (*a*) 7   (*b*) 9
   (*c*) 14   (*d*) 16

7 (*a*) characteristic: 2   (*b*) characteristic: 11
        mantissa: 0.1414   mantissa: 0.1345
   (*c*) characteristic: −2   (*d*) characteristic: −5
        mantissa: 0.7124   mantissa: 0.8814

8 (*a*) 2.8537   (*b*) 8.8762 − 10
   (*c*) 0.1271   (*d*) 1.3375
   (*e*) 8.4978 − 10   (*f*) 6.4978 − 10

9 (*a*) 940   (*b*) 0.973
   (*c*) 0.008512   (*d*) 4875

10 (*a*) 296,300   (*b*) 0.02465
   (*c*) 6.351   (*d*) 4.208
   (*e*) 26.72   (*f*) 0.4263
   (*g*) 0.5606   (*h*) 280.5

11 (*a*) $\log_2 14 = \dfrac{\ln 14}{\ln 2}$ (*b*) $\dfrac{\ln 88}{\ln 10}$

   (*c*) $\dfrac{\ln 75}{\ln 10}$ (*d*) $\dfrac{\ln y}{\ln x}$

   (*e*) $\dfrac{\ln a}{\ln 5}$ (*f*) $\dfrac{\ln 57}{\ln 4}$

12 (*a*) $x = 4$ (*b*) $x = 3$
   (*c*) $x = 2.6764$ (*d*) $x = 4.1390$

13 (*a*) SS = {25}   (*b*) SS = {3}
   (*c*) SS = ∅   (*d*) SS = ∅

Appendix
Tables

C

Table C-1

Natural
trigonometric
functions

Degrees	Sin	Cos	Tan	Cot	Sec	Csc	
0° 00′	.0000	1.0000	.0000		1.000		**90° 00′**
10	.0029	1.0000	.0029	343.8	1.000	343.8	50
20	.0058	1.0000	.0058	171.9	1.000	171.9	40
30	.0087	1.0000	.0087	114.6	1.000	114.6	30
40	.0116	.9999	.0116	85.94	1.000	85.95	20
50	.0145	.9999	.0145	68.75	1.000	68.76	10
1° 00′	.0175	.9998	.0175	57.29	1.000	57.30	**89° 00′**
10	.0204	.9998	.0204	49.10	1.000	49.11	50
20	.0233	.9997	.0233	42.96	1.000	42.98	40
30	.0262	.9997	.0262	38.19	1.000	38.20	30
40	.0291	.9996	.0291	34.37	1.000	34.38	20
50	.0320	.9995	.0320	31.24	1.001	31.26	10
2° 00′	.0349	.9994	.0349	28.64	1.001	28.65	**88° 00′**
10	.0378	.9993	.0378	26.43	1.001	26.45	50
20	.0407	.9992	.0407	24.54	1.001	24.56	40
30	.0436	.9990	.0437	22.90	1.001	22.93	30
40	.0465	.9989	.0466	21.47	1.001	21.49	20
50	.0494	.9988	.0495	20.21	1.001	20.23	10
3° 00′	.0523	.9986	.0524	19.08	1.001	19.11	**87° 00′**
10	.0552	.9985	.0553	18.07	1.002	18.10	50
20	.0581	.9983	.0582	17.17	1.002	17.20	40
30	.0610	.9981	.0612	16.35	1.002	16.38	30
40	.0640	.9980	.0641	15.60	1.002	15.64	20
50	.0669	.9978	.0670	14.92	1.002	14.96	10
4° 00′	.0698	.9976	.0699	14.30	1.002	14.34	**86° 00′**
10	.0727	.9974	.0729	13.73	1.003	13.76	50
20	.0756	.9971	.0758	13.20	1.003	13.23	40
30	.0785	.9969	.0787	12.71	1.003	12.75	30
40	.0814	.9967	.0816	12.25	1.003	12.29	20
50	.0843	.9964	.0846	11.83	1.004	11.87	10
5° 00′	.0872	.9962	.0875	11.43	1.004	11.47	**85° 00′**
10	.0901	.9959	.0904	11.06	1.004	11.10	50
20	.0929	.9957	.0934	10.71	1.004	10.76	40
30	.0958	.9954	.0963	10.39	1.005	10.43	30
40	.0987	.9951	.0992	10.08	1.005	10.13	20
50	.1016	.9948	.1022	9.788	1.005	9.839	10
6° 00′	.1045	.9945	.1051	9.514	1.006	9.567	**84° 00′**
10	.1074	.9942	.1080	9.255	1.006	9.309	50
20	.1103	.9939	.1110	9.010	1.006	9.065	40
30	.1132	.9936	.1139	8.777	1.006	8.834	30
40	.1161	.9932	.1169	8.556	1.007	8.614	20
50	.1190	.9929	.1198	8.345	1.007	8.405	10
7° 00′	.1219	.9925	.1228	8.144	1.008	8.206	**83° 00′**
10	.1248	.9922	.1257	7.953	1.008	8.016	50
20	.1276	.9918	.1287	7.770	1.008	7.834	40
30	.1305	.9914	.1317	7.596	1.009	7.661	30
40	.1334	.9911	.1346	7.429	1.009	7.496	20
50	.1363	.9907	.1376	7.269	1.009	7.337	10
8° 00′	.1392	.9903	.1405	7.115	1.010	7.185	**82° 00′**
10	.1421	.9899	.1435	6.968	1.010	7.040	50
20	.1449	.9894	.1465	6.827	1.011	6.900	40
30	.1478	.9890	.1495	6.691	1.011	6.765	30
40	.1507	.9886	.1524	6.561	1.012	6.636	20
50	.1536	.9881	.1554	6.435	1.012	6.512	10
9° 00′	.1564	.9877	.1584	6.314	1.012	6.392	**81° 00′**
	Cos	Sin	Cot	Tan	Csc	Sec	Degrees

SOURCE: From Gordon Fuller, *Plane Trigonometry with Tables*, 4th ed., McGraw-Hill, New York, 1972. Reprinted by permission of the publisher.

Table C-1

Natural
trigonometric
functions
(continued)

Degrees	Sin	Cos	Tan	Cot	Sec	Csc	
9° 00′	.1564	.9877	.1584	6.314	1.012	6.392	**81° 00′**
10	.1593	.9872	.1614	6.197	1.013	6.277	50
20	.1622	.9868	.1644	6.084	1.013	6.166	40
30	.1650	.9863	.1673	5.976	1.014	6.059	30
40	.1679	.9858	.1703	5.871	1.014	5.955	20
50	.1708	.9853	.1733	5.769	1.015	5.855	10
10° 00′	.1736	.9848	.1763	5.671	1.015	5.759	**80° 00′**
10	.1765	.9843	.1793	5.576	1.016	5.665	50
20	.1794	.9838	.1823	5.485	1.016	5.575	40
30	.1822	.9833	.1853	5.396	1.017	5.487	30
40	.1851	.9827	.1883	5.309	1.018	5.403	20
50	.1880	.9822	.1914	5.226	1.018	5.320	10
11° 00′	.1908	.9816	.1944	5.145	1.019	5.241	**79° 00′**
10	.1937	.9811	.1974	5.066	1.019	5.164	50
20	.1965	.9805	.2004	4.989	1.020	5.089	40
30	.1994	.9799	.2035	4.915	1.020	5.016	30
40	.2022	.9793	.2065	4.843	1.021	4.945	20
50	.2051	.9787	.2095	4.773	1.022	4.876	10
12° 00′	.2079	.9781	.2126	4.705	1.022	4.810	**78° 00′**
10	.2108	.9775	.2156	4.638	1.023	4.745	50
20	.2136	.9769	.2186	4.574	1.024	4.682	40
30	.2164	.9763	.2217	4.511	1.024	4.620	30
40	.2193	.9757	.2247	4.449	1.025	4.560	20
50	.2221	.9750	.2278	4.390	1.026	4.502	10
13° 00′	.2250	.9744	.2309	4.331	1.026	4.445	**77° 00′**
10	.2278	.9737	.2339	4.275	1.027	4.390	50
20	.2306	.9730	.2370	4.219	1.028	4.336	40
30	.2334	.9724	.2401	4.165	1.028	4.284	30
40	.2363	.9717	.2432	4.113	1.029	4.232	20
50	.2391	.9710	.2462	4.061	1.030	4.182	10
14° 00′	.2419	.9703	.2493	4.011	1.031	4.134	**76° 00′**
10	.2447	.9696	.2524	3.962	1.031	4.086	50
20	.2476	.9689	.2555	3.914	1.032	4.039	40
30	.2504	.9681	.2586	3.867	1.033	3.994	30
40	.2532	.9674	.2617	3.821	1.034	3.950	20
50	.2560	.9667	.2648	3.776	1.034	3.906	10
15° 00′	.2588	.9659	.2679	3.732	1.035	3.864	**75° 00′**
10	.2616	.9652	.2711	3.689	1.036	3.822	50
20	.2644	.9644	.2742	3.647	1.037	3.782	40
30	.2672	.9636	.2773	3.606	1.038	3.742	30
40	.2700	.9628	.2805	3.566	1.039	3.703	20
50	.2728	.9621	.2836	3.526	1.039	3.665	10
16° 00′	.2756	.9613	.2867	3.487	1.040	3.628	**74° 00′**
10	.2784	.9605	.2899	3.450	1.041	3.592	50
20	.2812	.9596	.2931	3.412	1.042	3.556	40
30	.2840	.9588	.2962	3.376	1.043	3.521	30
40	.2868	.9580	.2994	3.340	1.044	3.487	20
50	.2896	.9572	.3026	3.305	1.045	3.453	10
17° 00′	.2924	.9563	.3057	3.271	1.046	3.420	**73° 00′**
10	.2952	.9555	.3089	3.237	1.047	3.388	50
20	.2979	.9546	.3121	3.204	1.048	3.356	40
30	.3007	.9537	.3153	3.172	1.049	3.326	30
40	.3035	.9528	.3185	3.140	1.049	3.295	20
50	.3062	.9520	.3217	3.108	1.050	3.265	10
18° 00′	.3090	.9511	.3249	3.078	1.051	3.236	**72° 00′**
	Cos	Sin	Cot	Tan	Csc	Sec	Degrees

Table C-1

Natural trigonometric functions (continued)

Degrees	Sin	Cos	Tan	Cot	Sec	Csc	
18° 00'	.3090	.9511	.3249	3.078	1.051	3.236	**72° 00'**
10	.3118	.9502	.3281	3.047	1.052	3.207	50
20	.3145	.9492	.3314	3.018	1.053	3.179	40
30	.3173	.9483	.3346	2.989	1.054	3.152	30
40	.3201	.9474	.3378	2.960	1.056	3.124	20
50	.3228	.9465	.3411	2.932	1.057	3.098	10
19° 00'	.3256	.9455	.3443	2.904	1.058	3.072	**71° 00'**
10	.3283	.9446	.3476	2.877	1.059	3.046	50
20	.3311	.9436	.3508	2.850	1.060	3.021	40
30	.3338	.9426	.3541	2.824	1.061	2.996	30
40	.3365	.9417	.3574	2.798	1.062	2.971	20
50	.3393	.9407	.3607	2.773	1.063	2.947	10
20° 00'	.3420	.9397	.3640	2.747	1.064	2.924	**70° 00'**
10	.3448	.9387	.3673	2.723	1.065	2.901	50
20	.3475	.9377	.3706	2.699	1.066	2.878	40
30	.3502	.9367	.3739	2.675	1.068	2.855	30
40	.3529	.9356	.3772	2.651	1.069	2.833	20
50	.3557	.9346	.3805	2.628	1.070	2.812	10
21° 00'	.3584	.9336	.3839	2.605	1.071	2.790	**69° 00'**
10	.3611	.9325	.3872	2.583	1.072	2.769	50
20	.3638	.9315	.3906	2.560	1.074	2.749	40
30	.3665	.9304	.3939	2.539	1.075	2.729	30
40	.3692	.9293	.3973	2.517	1.076	2.709	20
50	.3719	.9283	.4006	2.496	1.077	2.689	10
22° 00'	.3746	.9272	.4040	2.475	1.079	2.669	**68° 00'**
10	.3773	.9261	.4074	2.455	1.080	2.650	50
20	.3800	.9250	.4108	2.434	1.081	2.632	40
30	.3827	.9239	.4142	2.414	1.082	2.613	30
40	.3854	.9228	.4176	2.394	1.084	2.595	20
50	.3881	.9216	.4210	2.375	1.085	2.577	10
23° 00'	.3907	.9205	.4245	2.356	1.086	2.559	**67° 00'**
10	.3934	.9194	.4279	2.337	1.088	2.542	50
20	.3961	.9182	.4314	2.318	1.089	2.525	40
30	.3987	.9171	.4348	2.300	1.090	2.508	30
40	.4014	.9159	.4383	2.282	1.092	2.491	20
50	.4041	.9147	.4417	2.264	1.093	2.475	10
24° 00'	.4067	.9135	.4452	2.246	1.095	2.459	**66° 00'**
10	.4094	.9124	.4487	2.229	1.096	2.443	50
20	.4120	.9112	.4522	2.211	1.097	2.427	40
30	.4147	.9100	.4557	2.194	1.099	2.411	30
40	.4173	.9088	.4592	2.177	1.100	2.396	20
50	.4200	.9075	.4628	2.161	1.102	2.381	10
25° 00'	.4226	.9063	.4663	2.145	1.103	2.366	**65° 00'**
10	.4253	.9051	.4699	2.128	1.105	2.352	50
20	.4279	.9038	.4734	2.112	1.106	2.337	40
30	.4305	.9026	.4770	2.097	1.108	2.323	30
40	.4331	.9013	.4806	2.081	1.109	2.309	20
50	.4358	.9001	.4841	2.066	1.111	2.295	10
26° 00'	.4384	.8988	.4877	2.050	1.113	2.281	**64° 00'**
10	.4410	.8975	.4913	2.035	1.114	2.268	50
20	.4436	.8962	.4950	2.020	1.116	2.254	40
30	.4462	.8949	.4986	2.006	1.117	2.241	30
40	.4488	.8936	.5022	1.991	1.119	2.228	20
50	.4514	.8923	.5059	1.977	1.121	2.215	10
27° 00'	.4540	.8910	.5095	1.963	1.122	2.203	**63° 00'**
	Cos	Sin	Cot	Tan	Csc	Sec	Degrees

Tables/**313**

Table C-1
Natural
trigonometric
functions
(continued)

Degrees	Sin	Cos	Tan	Cot	Sec	Csc	
27° 00′	.4540	.8910	.5095	1.963	1.122	2.203	**63° 00′**
10	.4566	.8897	.5132	1.949	1.124	2.190	50
20	.4592	.8884	.5169	1.935	1.126	2.178	40
30	.4617	.8870	.5206	1.921	1.127	2.166	30
40	.4643	.8857	.5243	1.907	1.129	2.154	20
50	.4669	.8843	5280	1.894	1.131	2.142	10
28° 00′	.4695	.8829	.5317	1.881	1.133	2.130	**62° 00′**
10	.4720	.8816	.5354	1.868	1.134	2.118	50
20	.4746	.8802	.5392	1.855	1.136	2.107	40
30	.4772	.8788	.5430	1.842	1.138	2.096	30
40	.4797	.8774	.5467	1.829	1.140	2.085	20
50	.4823	.8760	.5505	1 816	1.142	2.074	10
29° 00′	.4848	.8746	.5543	1.804	1.143	2.063	**61° 00′**
10	.4874	.8732	.5581	1.792	1.145	2.052	50
20	.4899	.8718	.5619	1.780	1.147	2.041	40
30	.4924	.8704	.5658	1.767	1.149	2.031	30
40	.4950	.8689	.5696	1.756	1.151	2.020	20
50	.4975	.8675	.5735	1.744	1.153	2.010	10
30° 00′	.5000	.8660	.5774	1.732	1.155	2.000	**60° 00′**
10	.5025	.8646	.5812	1.720	1.157	1.990	50
20	.5050	.8631	.5851	1.709	1.159	1.980	40
30	.5075	.8616	.5890	1.698	1.161	1.970	30
40	.5100	.8601	.5930	1.686	1.163	1.961	20
50	.5125	.8587	.5969	1.675	1.165	1.951	10
31° 00′	.5150	.8572	.6009	1.664	1.167	1.942	**59° 00′**
10	.5175	.8557	.6048	1.653	1.169	1.932	50
20	.5200	.8542	.6088	1.643	1.171	1.923	40
30	.5225	.8526	.6128	1.632	1.173	1.914	30
40	.5250	.8511	.6168	1.621	1.175	1.905	20
50	.5275	.8496	.6208	1.611	1.177	1.896	10
32° 00′	.5299	.8480	.6249	1.600	1.179	1.887	**58° 00′**
10	.5324	.8465	.6289	1.590	1.181	1.878	50
20	.5348	.8450	.6330	1.580	1.184	1.870	40
30	.5373	.8434	.6371	1.570	1.186	1.861	30
40	.5398	.8418	.6412	1.560	1.188	1.853	20
50	.5422	.8403	.6453	1.550	1.190	1.844	10
33° 00′	.5446	.8387	.6494	1.540	1.192	1.836	**57° 00′**
10	.5471	.8371	.6536	1.530	1.195	1.828	50
20	.5495	.8355	.6577	1.520	1.197	1.820	40
30	.5519	.8339	.6619	1.511	1.199	1.812	30
40	.5544	.8323	.6661	1.501	1.202	1.804	20
50	.5568	.8307	.6703	1.492	1.204	1.796	10
34° 00′	.5592	.8290	.6745	1.483	1.206	1.788	**56° 00′**
10	.5616	.8274	.6787	1.473	1.209	1.781	50
20	.5640	.8258	.6830	1.464	1.211	1.773	40
30	.5664	.8241	.6873	1.455	1.213	1.766	30
40	.5688	.8225	.6916	1.446	1.216	1.758	20
50	.5712	.8208	.6959	1.437	1.218	1.751	10
35° 00′	.5736	.8192	.7002	1.428	1.221	1.743	**55° 00′**
10	.5760	.8175	.7046	1.419	1.223	1.736	50
20	.5783	.8158	.7089	1.411	1.226	1.729	40
30	.5807	.8141	.7133	1.402	1.228	1.722	30
40	.5831	.8124	.7177	1.393	1.231	1.715	20
50	.5854	.8107	.7221	1.385	1.233	1.708	10
36° 00′	.5878	.8090	.7265	1.376	1.236	1.701	**54° 00′**
	Cos	Sin	Cot	Tan	Csc	Sec	Degrees

Table C-1

Natural
trigonometric
functions
(continued)

Degrees	Sin	Cos	Tan	Cot	Sec	Csc	
36° 00′	.5878	.8090	.7265	1.376	1.236	1.701	**54° 00′**
10	.5901	.8073	.7310	1.368	1.239	1.695	50
20	.5925	.8056	.7355	1.360	1.241	1.688	40
30	.5948	.8039	.7400	1.351	1.244	1.681	30
40	.5972	.8021	.7445	1.343	1.247	1.675	20
50	.5995	.8004	.7490	1.335	1.249	1.668	10
37° 00′	.6018	.7986	.7536	1.327	1.252	1.662	**53° 00′**
10	.6041	.7969	.7581	1.319	1.255	1.655	50
20	.6065	:7951	.7627	1.311	1.258	1.649	40
30	.6088	.7934	.7673	1.303	1.260	1.643	30
40	.6111	.7916	.7720	1.295	1.263	1.636	20
50	.6134	.7898	.7766	1.288	1.266	1.630	10
38° 00′	.6157	.7880	.7813	1.280	1.269	1.624	**52° 00′**
10	.6180	.7862	.7860	1.272	1.272	1.618	50
20	.6202	.7844	.7907	1.265	1.275	1.612	40
30	.6225	.7826	.7954	1.257	1.278	1.606	30
40	.6248	.7808	.8002	1.250	1.281	1.601	20
50	.6271	.7790	.8050	1.242	1.284	1.595	10
39° 00′	.6293	.7771	.8098	1.235	1.287	1.589	**51° 00′**
10	.6316	.7753	.8146	1.228	1.290	1.583	50
20	.6338	.7735	.8195	1.220	1.293	1.578	40
30	.6361	.7716	.8243	1.213	1.296	1.572	30
40	.6383	.7698	.8292	1.206	1.299	1.567	20
50	.6406	.7679	.8342	1.199	1.302	1.561	10
40° 00′	.6428	.7660	.8391	1.192	1.305	1.556	**50° 00′**
10	.6450	.7642	.8441	1.185	1.309	1.550	50
20	.6472	.7623	.8491	1.178	1.312	1.545	40
30	.6494	.7604	.8541	1.171	1.315	1.540	30
40	.6517	.7585	.8591	1.164	1.318	1.535	20
50	.6539	.7566	.8642	1.157	1.322	1.529	10
41° 00′	.6561	.7547	.8693	1.150	1.325	1.524	**49° 00′**
10	.6583	.7528	.8744	1.144	1.328	1.519	50
20	.6604	.7509	.8796	1.137	1.332	1.514	40
30	.6626	.7490	.8847	1.130	1.335	1.509	30
40	.6648	.7470	.8899	1.124	1.339	1.504	20
50	.6670	.7451	.8952	1.117	1.342	1.499	10
42° 00′	.6691	.7431	.9004	1.111	1.346	1.494	**48° 00′**
10	.6713	.7412	.9057	1.104	1.349	1.490	50
20	.6734	.7392	.9110	1.098	1.353	1.485	40
30	.6756	.7373	.9163	1.091	1.356	1.480	30
40	.6777	.7353	.9217	1·085	1.360	1.476	20
50	.6799	.7333	.9271	1.079	1.364	1.471	10
43° 00′	.6820	.7314	.9325	1.072	1.367	1.466	**47° 00′**
10	.6841	.7294	.9380	1.066	1.371	1.462	50
20	.6862	.7274	.9435	1.060	1.375	1.457	40
30	.6884	.7254	.9490	1.054	1.379	1.453	30
40	.6905	.7234	.9545	1.048	1.382	1.448	20
50	.6926	.7214	.9601	1.042	1.386	1.444	10
44° 00′	.6947	.7193	.9657	1.036	1.390	1.440	**46° 00′**
10	.6967	.7173	.9713	1.030	1.394	1.435	50
20	.6988	.7153	.9770	1.024	1.398	1.431	40
30	.7009	.7133	.9827	1.018	1.402	1.427	30
40	.7030	.7112	.9884	1.012	1.406	1.423	20
50	.7050	.7092	.9942	1.006	1.410	1.418	10
45° 00′	.7071	.7071	1.000	1.000	1.414	1.414	**45° 00′**
	Cos	Sin	Cot	Tan	Csc	Sec	Degrees

Table C-2

Trigonometric ratios for angles measured in radians at intervals of 0.01 radian, from 0.00 to 1.60

t	$\sin t$	$\cos t$	$\tan t$	$\cot t$	$\sec t$	$\csc t$
0.00	0.0000	1.000	0.0000		1.000	
0.01	0.01000	1.000	0.01000	100.0	1.000	100.0
0.02	0.02000	0.9998	0.02000	49.99	1.000	50.00
0.03	0.03000	0.9996	0.03001	33.32	1.000	33.34
0.04	0.03999	0.9992	0.04002	24.99	1.001	25.01
0.05	0.04998	0.9988	0.05004	19.98	1.001	20.01
0.06	0.06000	0.9982	0.06007	16.65	1.002	16.68
0.07	0.06994	0.9976	0.07011	14.26	1.002	14.30
0.08	0.07991	0.9968	0.08017	12.47	1.003	12.51
0.09	0.08988	0.9960	0.09024	11.08	1.004	11.13
0.10	0.09983	0.9950	0.1003	9.967	1.005	10.02
0.11	0.1098	0.9940	0.1104	9.054	1.006	9.109
0.12	0.1197	0.9928	0.1206	8.293	1.007	8.353
0.13	0.1296	0.9916	0.1307	7.649	1.009	7.714
0.14	0.1395	0.9902	0.1409	7.096	1.010	7.166
0.15	0.1494	0.9888	0.1511	6.617	1.011	6.692
0.16	0.1593	0.9872	0.1614	6.197	1.013	6.277
0.17	0.1692	0.9856	0.1717	5.826	1.015	5.911
0.18	0.1790	0.9838	0.1820	5.495	1.016	5.586
0.19	0.1889	0.9820	0.1923	5.200	1.018	5.295
0.20	0.1987	0.9801	0.2027	4.933	1.020	5.033
0.21	0.2085	0.9780	0.2131	4.692	1.022	4.797
0.22	0.2182	0.9759	0.2236	4.472	1.025	4.582
0.23	0.2280	0.9737	0.2341	4.271	1.027	4.386
0.24	0.2377	0.9713	0.2447	4.086	1.030	4.207
0.25	0.2474	0.9689	0.2553	3.916	1.032	4.042
0.26	0.2571	0.9664	0.2660	3.759	1.035	3.890
0.27	0.2667	0.9638	0.2768	3.613	1.038	3.749
0.28	0.2764	0.9611	0.2876	3.478	1.041	3.619
0.29	0.2860	0.9582	0.2984	3.351	1.044	3.497
0.30	0.2955	0.9553	0.3093	3.233	1.047	3.384
0.31	0.3051	0.9523	0.3203	3.122	1.050	3.278
0.32	0.3146	0.9492	0.3314	3.018	1.053	3.179
0.33	0.3240	0.9460	0.3425	2.920	1.057	3.086
0.34	0.3335	0.9428	0.3537	2.827	1.061	2.999
0.35	0.3429	0.9394	0.3650	2.740	1.065	2.916
0.36	0.3523	0.9359	0.3764	2.657	1.068	2.839
0.37	0.3616	0.9323	0.3879	2.578	1.073	2.765
0.38	0.3709	0.9287	0.3994	2.504	1.077	2.696
0.39	0.3802	0.9249	0.4111	2.433	1.081	2.630
t	$\sin t$	$\cos t$	$\tan t$	$\cot t$	$\sec t$	$\csc t$

SOURCE: From Howard Taylor and Thomas L. Wade, *Contemporary Trigonometry*, McGraw-Hill, New York, 1973. Reprinted by permission of the publisher.

Table C-2

Trigonometric ratios for angles measured in radians at intervals of 0.01 radian, from 0.00 to 1.60

t	sin t	cos t	tan t	cot t	sec t	csc
0.40	0.3894	0.9211	0.4228	2.365	1.086	2.5
0.41	0.3986	0.9171	0.4346	2.301	1.090	2.5
0.42	0.4078	0.9131	0.4466	2.239	1.095	2.4
0.43	0.4169	0.9090	0.4586	2.180	1.100	2.3
0.44	0.4259	0.9048	0.4708	2.124	1.105	2.3
0.45	0.4350	0.9004	0.4831	2.070	1.111	2.2
0.46	0.4439	0.8961	0.4954	2.018	1.116	2.2
0.47	0.4529	0.8916	0.5080	1.969	1.122	2.2
0.48	0.4618	0.8870	0.5206	1.921	1.127	2.1
0.49	0.4706	0.8823	0.5334	1.875	1.133	2.1
0.50	0.4794	0.8776	0.5463	1.830	1.139	2.0
0.51	0.4882	0.8727	0.5594	1.788	1.146	2.0
0.52	0.4969	0.8678	0.5726	1.747	1.152	2.0
0.53	0.5055	0.8628	0.5859	1.707	1.159	1.9
0.54	0.5141	0.8577	0.5994	1.668	1.166	1.9
0.55	0.5227	0.8525	0.6131	1.631	1.173	1.9
0.56	0.5312	0.8473	0.6269	1.595	1.180	1.88
0.57	0.5396	0.8419	0.6410	1.560	1.188	1.85
0.58	0.5480	0.8365	0.6552	1.526	1.196	1.82
0.59	0.5564	0.8309	0.6696	1.494	1.203	1.79
0.60	0.5646	0.8253	0.6841	1.462	1.212	1.77
0.61	0.5729	0.8196	0.6989	1.431	1.220	1.74
0.62	0.5810	0.8139	0.7139	1.401	1.229	1.72
0.63	0.5891	0.8080	0.7291	1.372	1.238	1.69
0.64	0.5972	0.8021	0.7445	1.343	1.247	1.67
0.65	0.6052	0.7961	0.7602	1.315	1.256	1.65
0.66	0.6131	0.7900	0.7761	1.288	1.266	1.63
0.67	0.6210	0.7838	0.7923	1.262	1.276	1.61
0.68	0.6288	0.7776	0.8087	1.237	1.286	1.59
0.69	0.6365	0.7712	0.8253	1.212	1.297	1.57
0.70	0.6442	0.7648	0.8423	1.187	1.307	1.55
0.71	0.6518	0.7584	0.8595	1.163	1.319	1.53
0.72	0.6594	0.7518	0.8771	1.140	1.330	1.51
0.73	0.6669	0.7452	0.8949	1.117	1.342	1.50
0.74	0.6743	0.7385	0.9131	1.095	1.354	1.48
0.75	0.6816	0.7317	0.9316	1.073	1.367	1.46
0.76	0.6889	0.7248	0.9505	1.052	1.380	1.45
0.77	0.6961	0.7179	0.9697	1.031	1.393	1.43
0.78	0.7033	0.7109	0.9893	1.011	1.407	1.42
0.79	0.7104	0.7038	1.009	0.9908	1.421	1.40
t	sin t	cos t	tan t	cot t	sec t	csc

Table C-2

Trigonometric ratios for angles measured in radians at intervals of 0.01 radian, from 0.00 to 1.60 (continued)

t	$\sin t$	$\cos t$	$\tan t$	$\cot t$	$\sec t$	$\csc t$
0.80	0.7174	0.6967	1.030	0.9712	1.435	1.394
0.81	0.7243	0.6895	1.050	0.9520	1.450	1.381
0.82	0.7311	0.6822	1.072	0.9331	1.466	1.368
0.83	0.7379	0.6749	1.093	0.9146	1.482	1.355
0.84	0.7446	0.6675	1.116	0.8964	1.498	1.343
0.85	0.7513	0.6600	1.138	0.8785	1.515	1.331
0.86	0.7578	0.6524	1.162	0.8609	1.533	1.320
0.87	0.7643	0.6448	1.185	0.8437	1.551	1.308
0.88	0.7707	0.6372	1.210	0.8267	1.569	1.297
0.89	0.7771	0.6294	1.235	0.8100	1.589	1.287
0.90	0.7833	0.6216	1.260	0.7936	1.609	1.277
0.91	0.7895	0.6137	1.286	0.7774	1.629	1.267
0.92	0.7956	0.6058	1.313	0.7615	1.651	1.257
0.93	0.8016	0.5978	1.341	0.7458	1.673	1.247
0.94	0.8076	0.5898	1.369	0.7303	1.696	1.238
0.95	0.8134	0.5817	1.398	0.7151	1.719	1.229
0.96	0.8192	0.5735	1.428	0.7001	1.744	1.221
0.97	0.8249	0.5653	1.459	0.6853	1.769	1.212
0.98	0.8305	0.5570	1.491	0.6707	1.795	1.204
0.99	0.8360	0.5487	1.524	0.6563	1.823	1.196
1.00	0.8415	0.5403	1.557	0.6421	1.851	1.188
1.01	0.8468	0.5319	1.592	0.6281	1.880	1.181
1.02	0.8521	0.5234	1.628	0.6142	1.911	1.174
1.03	0.8573	0.5148	1.665	0.6005	1.942	1.166
1.04	0.8624	0.5062	1.704	0.5870	1.975	1.160
1.05	0.8674	0.4976	1.743	0.5736	2.010	1.153
1.06	0.8724	0.4889	1.784	0.5604	2.046	1.146
1.07	0.8772	0.4801	1.827	0.5473	2.083	1.140
1.08	0.8820	0.4713	1.871	0.5344	2.122	1.134
1.09	0.8866	0.4625	1.917	0.5216	2.162	1.128
1.10	0.8912	0.4536	1.965	0.5090	2.205	1.122
1.11	0.8957	0.4447	2.014	0.4964	2.249	1.116
1.12	0.9001	0.4357	2.066	0.4840	2.295	1.111
1.13	0.9044	0.4267	2.120	0.4718	2.344	1.106
1.14	0.9086	0.4176	2.176	0.4596	2.395	1.101
1.15	0.9128	0.4085	2.234	0.4475	2.448	1.096
1.16	0.9168	0.3993	2.296	0.4356	2.504	1.091
1.17	0.9208	0.3902	2.360	0.4237	2.563	1.086
1.18	0.9246	0.3809	2.427	0.4120	2.625	1.082
1.19	0.9284	0.3717	2.498	0.4003	2.691	1.077
t	$\sin t$	$\cos t$	$\tan t$	$\cot t$	$\sec t$	$\csc t$

Table C-2

Trigonometric ratios for angles measured in radians at intervals of 0.01 radian, from 0.00 to 1.60 (continue

t	sin t	cos t	tan t	cot t	sec t	csc
1.20	0.9320	0.3624	2.572	0.3888	2.760	1.0
1.21	0.9356	0.3530	2.650	0.3773	2.833	1.0
1.22	0.9391	0.3436	2.733	0.3659	2.910	1.0
1.23	0.9425	0.3342	2.820	0.3546	2.992	1.0
1.24	0.9458	0.3248	2.912	0.3434	3.079	1.0
1.25	0.9490	0.3153	3.010	0.3323	3.171	1.0
1.26	0.9521	0.3058	3.113	0.3212	3.270	1.0
1.27	0.9551	0.2963	3.224	0.3102	3.375	1.0
1.28	0.9580	0.2867	2.341	0.2993	3.488	1.0
1.29	0.9608	0.2771	3.467	0.2884	3.609	1.0
1.30	0.9636	0.2675	3.602	0.2776	3.738	1.0
1.31	0.9662	0.2579	3.747	0.2669	3.878	1.0
1.32	0.9687	0.2482	3.903	0.2562	4.029	1.0
1.33	0.9711	0.2385	4.072	0.2456	4.193	1.0
1.34	0.9735	0.2288	4.256	0.2350	4.372	1.0
1.35	0.9757	0.2190	4.455	0.2245	4.566	1.0
1.36	0.9779	0.2092	4.673	0.2140	4.779	1.0
1.37	0.9799	0.1994	4.913	0.2035	5.014	1.0
1.38	0.9819	0.1896	5.177	0.1931	5.273	1.01
1.39	0.9837	0.1798	5.471	0.1828	5.561	1.01
1.40	0.9854	0.1700	5.798	0.1725	5.883	1.01
1.41	0.9871	0.1601	6.165	0.1622	6.246	1.01
1.42	0.9887	0.1502	6.581	0.1519	6.657	1.01
1.43	0.9901	0.1403	7.055	0.1417	7.126	1.01
1.44	0.9915	0.1304	7.602	0.1315	7.667	1.00
1.45	0.9927	0.1205	8.238	0.1214	8.299	1.00
1.46	0.9939	0.1106	8.989	0.1113	9.044	1.00
1.47	0.9949	0.1006	9.887	0.1011	9.938	1.00
1.48	0.9959	0.09067	10.98	0.09105	11.03	1.00
1.49	0.9967	0.08071	12.35	0.08097	12.39	1.00
1.50	0.9975	0.07074	14.10	0.07091	14.14	1.00
1.51	0.9982	0.06076	16.43	0.06087	16.46	1.00
1.52	0.9987	0.05077	19.67	0.05084	19.70	1.00
1.53	0.9992	0.04079	24.50	0.04082	24.52	1.00
1.54	0.9995	0.03079	32.46	0.03081	32.48	1.00
1.55	0.9998	0.02079	48.08	0.02080	48.09	1.00
1.56	0.9999	0.01080	92.62	0.01080	92.63	1.00
1.57	1.000	0.00080	1256	0.00080	1256	1.00
1.58	1.000	−0.00920	−108.7	−0.00920	−108.7	1.00
1.59	0.9998	−0.01920	−52.07	−0.01921	−52.08	1.00
1.60	0.9996	−0.02920	−34.23	−0.02921	−34.25	1.00
t	sin t	cos t	tan t	cot t	sec t	csc

Table C-3
Logarithms of numbers

N	0	1	2	3	4	5	6	7	8	9
10	0000	0043	0086	0128	0170	0212	0253	0294	0334	0374
11	0414	0453	0492	0531	0569	0607	0645	0682	0719	0755
12	0792	0828	0864	0899	0934	0969	1004	1038	1072	1106
13	1139	1173	1206	1239	1271	1303	1335	1367	1399	1430
14	1461	1492	1523	1553	1584	1614	1644	1673	1703	1732
15	1761	1790	1818	1847	1875	1903	1931	1959	1987	2014
16	2041	2068	2095	2122	2148	2175	2201	2227	2253	2279
17	2304	2330	2355	2380	2405	2430	2455	2480	2504	2529
18	2553	2577	2601	2625	2648	2672	2695	2718	2742	2765
19	2788	2810	2833	2856	2878	2900	2923	2945	2967	2989
20	3010	3032	3054	3075	3096	3118	3139	3160	3181	3201
21	3222	3243	3263	3284	3304	3324	3345	3365	3385	3404
22	3424	3444	3464	3483	3502	3522	3541	3560	3579	3598
23	3617	3636	3655	3674	3692	3711	3729	3747	3766	3784
24	3802	3820	3838	3856	3874	3892	3909	3927	3945	3962
25	3979	3997	4014	4031	4048	4065	4082	4099	4116	4133
26	4150	4166	4183	4200	4216	4232	4249	4265	4281	4298
27	4314	4330	4346	4362	4378	4393	4409	4425	4440	4456
28	4472	4487	4502	4518	4533	4548	4564	4579	4594	4609
29	4624	4639	4654	4669	4683	4698	4713	4728	4742	4757
30	4771	4786	4800	4814	4829	4843	4857	4871	4886	4900
31	4914	4928	4942	4955	4969	4983	4997	5011	5024	5038
32	5051	5065	5079	5092	5105	5119	5132	5145	5159	5172
33	5185	5198	5211	5224	5237	5250	5263	5276	5289	5302
34	5315	5328	5340	5353	5366	5378	5391	5403	5416	5428
35	5441	5453	5465	5478	5490	5502	5514	5527	5539	5551
36	5563	5575	5587	5599	5611	5623	5635	5647	5658	5670
37	5682	5694	5705	5717	5729	5740	5752	5763	5775	5786
38	5798	5809	5821	5832	5843	5855	5866	5877	5888	5899
39	5911	5922	5933	5944	5955	5966	5977	5988	5999	6010
40	6021	6031	6042	6053	6064	6075	6085	6096	6107	6117
41	6128	6138	6149	6160	6170	6180	6191	6201	6212	6222
42	6232	6243	6253	6263	6274	6284	6294	6304	6314	6325
43	6335	6345	6355	6365	6375	6385	6395	6405	6415	6425
44	6435	6444	6454	6464	6474	6484	6493	6503	6513	6522
45	6532	6542	6551	6561	6571	6580	6590	6599	6609	6618
46	6628	6637	6646	6656	6665	6675	6684	6693	6702	6712
47	6721	6730	6739	6749	6758	6767	6776	6785	6794	6803
48	6812	6821	6830	6839	6848	6857	6866	6875	6884	6893
49	6902	6911	6920	6928	6937	6946	6955	6964	6972	6981
50	6990	6998	7007	7016	7024	7033	7042	7050	7059	7067
51	7076	7084	7093	7101	7110	7118	7126	7135	7143	7152
52	7160	7168	7177	7185	7193	7202	7210	7218	7226	7235
52	7243	7251	7259	7267	7275	7284	7292	7300	7308	7316
54	7324	7332	7340	7348	7356	7364	7372	7380	7388	7396
N	0	1	2	3	4	5	6	7	8	9

Table C-3

Logarithms of numbers (continued)

N	0	1	2	3	4	5	6	7	8	9
55	7404	7412	7419	7427	7435	7443	7451	7459	7466	7474
56	7482	7490	7497	7505	7513	7520	7528	7536	7543	7551
57	7559	7566	7574	7582	7589	7597	7604	7612	7619	7627
58	7634	7642	7649	7657	7664	7672	7679	7686	7694	7701
59	7709	7716	7723	7731	7738	7745	7752	7760	7767	7774
60	7782	7789	7796	7803	7810	7818	7825	7832	7839	7846
61	7853	7860	7868	7875	7882	7889	7896	7903	7910	7917
62	7924	7931	7938	7945	7952	7959	7966	7973	7980	7987
63	7993	8000	8007	8014	8021	8028	8035	8041	8048	8055
64	8062	8069	8075	8082	8089	8096	8102	8109	8116	8122
65	8129	8136	8142	8149	8156	8162	8169	8176	8182	8189
66	8195	8202	8209	8215	8222	8228	8235	8241	8248	8254
67	8261	8267	8274	8280	8287	8293	8299	8306	8312	8319
68	8325	8331	8338	8344	8351	8357	8363	8370	8376	8382
69	8388	8395	8401	8407	8414	8420	8426	8432	8439	8445
70	8451	8457	8463	8470	8476	8482	8488	8494	8500	8506
71	8513	8519	8525	8531	8537	8543	8549	8555	8561	8567
72	8573	8579	8585	8591	8597	8603	8609	8615	8621	8627
73	8633	8639	8645	8651	8657	8663	8669	8675	8681	8686
74	8692	8698	8704	8710	8716	8722	8727	8733	8739	8745
75	8751	8756	8762	8768	8774	8779	8785	8791	8797	8802
76	8808	8814	8820	8825	8831	8837	8842	8848	8854	8859
77	8865	8871	8876	8882	8887	8893	8899	8904	8910	8915
78	8921	8927	8932	8938	8943	8949	8954	8960	8965	8971
79	8976	8982	8987	8993	8998	9004	9009	9015	9020	9025
80	9031	9036	9042	9047	9053	9058	9063	9069	9074	9079
81	9085	9090	9096	9101	9106	9112	9117	9122	9128	9133
82	9138	9143	9149	9154	9159	9165	9170	9175	9180	9186
83	9191	9196	9201	9206	9212	9217	9222	9227	9232	9238
84	9243	9248	9253	9258	9263	9269	9274	9279	9284	9289
85	9294	9299	9304	9309	9315	9320	9325	9330	9335	9340
86	9345	9350	9355	9360	9365	9370	9375	9380	9385	9390
87	9395	9400	9405	9410	9415	9420	9425	9430	9435	9440
88	9445	9450	9455	9460	9465	9469	9474	9479	9484	9489
89	9494	9499	9504	9509	9513	9518	9523	9528	9533	9538
90	9542	9547	9552	9557	9562	9566	9571	9576	9581	9586
91	9590	9595	9600	9605	9609	9614	9619	9624	9628	9633
92	9638	9643	9647	9652	9657	9661	9666	9671	9675	9680
93	9685	9689	9694	9699	9703	9708	9713	9717	9722	9727
94	9731	9736	9741	9745	9750	9754	9759	9763	9768	9773
95	9777	9782	9786	9791	9795	9800	9805	9809	9814	9818
96	9823	9827	9832	9836	9841	9845	9850	9854	9859	9863
97	9868	9872	9877	9881	9886	9890	9894	9899	9903	9908
98	9912	9917	9921	9926	9930	9934	9939	9943	9948	9952
99	9956	9961	9965	9969	9974	9978	9983	9987	9991	9996
N	0	1	2	3	4	5	6	7	8	9

SOURCE: From Gordon Fuller, *Plane Trigonometry with Tables*, 4th ed., McGraw-Hill, New York, 1972. Reprinted by permission of the publisher.

Table C-4

Powers
and roots

No.	Sq.	Sq. root	Cube	Cube root	No.	Sq.	Sq. root	Cube	Cube root
1	1	1.000	1	1.000	51	2,601	7.141	132,651	3.708
2	4	1.414	8	1.260	52	2,704	7.211	140,608	3.733
3	9	1.732	27	1.442	53	2,809	7.280	148,877	3.756
4	16	2.000	64	1.587	54	2,916	7.348	157,464	3.780
5	25	2.236	125	1.710	55	3,025	7.416	166,375	3.803
6	36	2.449	216	1.817	56	3,136	7.483	175,616	3.826
7	49	2.646	343	1.913	57	3,249	7.550	185,193	3.849
8	64	2.828	512	2.000	58	3,364	7.616	195,112	3.871
9	81	3.000	729	2.080	59	3,481	7.681	205,379	3.893
10	100	3.162	1,000	2.154	60	3,600	7.746	216,000	3.915
11	121	3.317	1,331	2.224	61	3,721	7.810	226,981	3.936
12	144	3.464	1,728	2.289	62	3,844	7.874	238,328	3.958
13	169	3.606	2,197	2.351	63	3,969	7.937	250,047	3.979
14	196	3.742	2,744	2.410	64	4,096	8.000	262,144	4.000
15	225	3.873	3,375	2.466	65	4,225	8.062	274,625	4.021
16	256	4.000	4,096	2.520	66	4,356	8.124	287,496	4.041
17	289	4.123	4,913	2.571	67	4,489	8.185	300,763	4.062
18	324	4.243	5,832	2.621	68	4,624	8.246	314,432	4.082
19	361	4.359	6,859	2.668	69	4,761	8.307	328,509	4.102
20	400	4.472	8,000	2.714	70	4,900	8.367	343,000	4.121
21	441	4.583	9,261	2.759	71	5,041	8.426	357,911	4.141
22	484	4.690	10,648	2.802	72	5,184	8.485	373,248	4.160
23	529	4.796	12,167	2.844	73	5,329	8.544	389,017	4.179
24	576	4.899	13,824	2.884	74	5,476	8.602	405,224	4.198
25	625	5.000	15,625	2.924	75	5,625	8.660	421,875	4.217
26	676	5.099	17,576	2.962	76	5,776	8.718	438,976	4.236
27	729	5.196	19,683	3.000	77	5,929	8.775	456,533	4.254
28	784	5.292	21,952	3.037	78	6,084	8.832	474,552	4.273
29	841	5.385	24,389	3.072	79	6,241	8.888	493,039	4.291
30	900	5.477	27,000	3.107	80	6,400	8.944	512,000	4.309
31	961	5.568	29,791	3.141	81	6,561	9.000	531,441	4.327
32	1,024	5.657	32,768	3.175	82	6,724	9.055	551,368	4.344
33	1,089	5.745	35,937	3.208	83	6,889	9.110	571,787	4.362
34	1,156	5.831	39,304	3.240	84	7,056	9.165	592,704	4.380
35	1,225	5.916	42,875	3.271	85	7,225	9.220	614,125	4.397
36	1,296	6.000	46,656	3.302	86	7,396	9.274	636,056	4.414
37	1,369	6.083	50,653	3.332	87	7,569	9.327	658,503	4.431
38	1,444	6.164	54,872	3.362	88	7,744	9.381	681,472	4.448
39	1,521	6.245	59,319	3.391	89	7,921	9.434	704,969	4.465
40	1,600	6.325	64,000	3.420	90	8,100	9.487	729,000	4.481
41	1,681	6.403	68,921	3.448	91	8,281	9.539	753,571	4.498
42	1,764	6.481	74,088	3.476	92	8,464	9.592	778,688	4.514
43	1,849	6.557	79,507	3.503	93	8,649	9.644	804,357	4.531
44	1,936	6.633	85,184	3.530	94	8,836	9.695	830,584	4.547
45	2,025	6.708	91,125	3.557	95	9,025	9.747	857,375	4.563
46	2,116	6.782	97,336	3.583	96	9,216	9.798	884,736	4.579
47	2,209	6.856	103,823	3.609	97	9,409	9.849	912,673	4.595
48	2,304	6.928	110,592	3.634	98	9,604	9.899	941,192	4.610
49	2,401	7.000	117,649	3.659	99	9,801	9.950	970,299	4.626
50	2,500	7.071	125,000	3.684	100	10,000	10.000	1,000,000	4.642

SOURCE: From Gordon Fuller, *Plane Trigonometry with Tables*, 4th ed., McGraw-Hill, New York, 1972. Reprinted by permission of the publisher.

Index